T0191688

Mathematical Problems in Data Science

Li M. Chen • Zhixun Su • Bo Jiang

Mathematical Problems in Data Science

Theoretical and Practical Methods

 Springer

Li M. Cheng
Department of Computer Science
    and Information Technology
The University of the District of Columbia
Washington, DC, USA

Zhixun Su
School of Mathematical Sciences
Dalian University of Technology
Dalian, China

Bo Jiang
School of Information Science
    and Technology
Dalian Maritime University
Dalian, China

ISBN 978-3-319-79739-7   ISBN 978-3-319-25127-1   (eBook)
DOI 10.1007/978-3-319-25127-1

Springer Cham Heidelberg New York Dordrecht London
© Springer International Publishing Switzerland 2015
Softcover re-print of the Hardcover 1st edition 2015

Printed on acid-free paper

Springer International Publishing AG Switzerland is part of Springer Science+Business Media (www.
springer.com)

To ICM 2014 and Its Satellite Conference on Data Science

Preface

Modern data science is related to massive data sets (BigData), machine learning, and cloud computing. There are multiple ways of understanding data science: (1) BigData with small cloud computational power, which requires very fast algorithms; (2) relatively small data sets with large cloud computational power, which we can compute by distributing data to the cloud without a very efficient algorithm; (3) BigData and with big cloud, which requires both techniques of algorithms and architectural infrastructures (new computing models); and (4) small data sets with small cloud, which just requires the standard methods.

This book contains state-of-the-art knowledge for researchers in data science. It also presents various problems in BigData and data science. We first introduce important statistical and computational methods for data analysis. For example, we discuss the principal component analysis for the dimension reduction of massive data sets. Then, we introduce graph theoretical methods such as GraphCut, the Laplacian matrix, and Google PageRank for data search and classification. We also discuss efficient algorithms, the hardness of problems involving various types of BigData, and geometric data structures. This book is particularly interested in the discussion of incomplete data sets and partial connectedness among data points or data sets. The second part of the book focuses on special topics, which cover topological analysis and machine learning, business and financial data recovery, and massive data classification and predication for high-dimensional data sets. Another purpose of this book is to challenge the major ongoing and unsolved problems in data science and provide some prospective solutions to these problems.

This book is a concise and quick introduction to the hottest topic in mathematics, computer science, and information technology today: data science. Data science first emerged in mathematics and computer science out of the research need for the numerous applications of BigData in the information technology, business, and medical industries. This book has two main objectives. The first objective of this book is to cover necessary knowledge in statistics, graph theory, algorithms, and computational science. There is also specific focus on the internal connectivity of incomplete data sets, which could be one of the central topics of future data science, unlike the existing method of data processing where data modeling is at the center.

The second focus of this book discusses major ongoing and unsolved problems in data science and provides some prospective solutions for these problems.

The book also collects some research papers from the talks given at the International Congress of Mathematics (ICM) 2014 Satellite Conference on Mathematical Foundation of Modern Data Sciences Computing, Logic, and Education, Dalian Maritime University, Dalian, China, which took place from July 27 to August 1, 2014. We are grateful to the Seoul ICM 2014 organization committee and National Science Foundation of China for their support. Many thanks go to Professor Reinhard Klette at the University of Auckland and Professor Wen Gao at Beijing University for giving excellent invited talks. Special thanks to Professors Shi-Qiang Wang (Beijing Normal University), Steven G. Krantz (Washington University), Shmuel Weinberger (University of Chicago), and Hanan Samet (University of Maryland) for their support. Special thanks also go to Dalian University of Technology, Dalian Maritime University, Southeast University of China, and University of the District of Columbia for their support to this conference.

This book has three parts. The first part contains the basics in data science; the second part mainly deals with computing, leaning, and problems in data science; the third part is selected topics. Chapter 1: Introduction (L. Chen); Chap. 2: Overview of Basic Methods for Data Science (L. Chen); Chap. 3: Relationship and Connectivity of Incomplete Data Collection (L. Chen); Chap. 4: Machine Leaning for Data Science (L. Chen); Chap. 5: Images, Videos, and BigData (L. Chen); Chap. 6: Topological Data Analysis (L. Chen); Chap. 7: Monte Carlo Methods and Their Applications in Big Data Analysis (H. Ji and Y. Li); Chap. 8: Feature Extraction via Vector Bundle Learning (R. Liu and Z. Su); Chap. 9: Curve Interpolation and Positivity-Preserving Financial Curve Construction (P. Huang, H. Wang, P. Wu, and Y. Li); Chap. 10: Advanced Methods in Variational Learning (J. Spencer and K. Chen); Chap. 11: On-line Strategies of Groups Evacuation from a Convex Region in the Plane (B. Jiang, Y. Liu, and H. Zhang); and Chap. 12: A New Computational Model of Bigdata (B. Zhu).

Washington, DC, USA   Li M. Chen
Dalian, China   Zhixun Su
                                                                                       Bo Jiang

Contents

Acronyms

N	The natural number set
I	The integer number set
R	The real number set
$G = (V, E)$	A graph G with the vertex set V and the edge set E
$|A|$	The number of elements in set A
$d(x, y)$	The distance between x and y
kNN	k nearest neighbor classification method
E_m	m_dimensional Euclidean space
Σ_m	m_dimensional digital space
B^k	k-dimensional unite ball
S^k	k-dimensional unite sphere
∂M	The boundary of M
$O(f)$	The time cost of an algorithm which is smaller than a constant $c \cdot f$
f'	The derivative of function f
MST	The minimum spanning tree
f_x	The partial derivative of function f with respect to the variable x
$\partial f / \partial x$	The same as f_x
TDA	Topological data analysis

Part I
Basic Data Science

Chapter 1
Introduction: Data Science and BigData Computing

Li M. Chen

Abstract What is Data Science? Data contains science. It is much different from the angle of classical mathematics that uses mathematical models to fit the data. Today, we are supposed to find rules and properties in the data set, even among different data sets. In this chapter, we will explain data science and its relationship to BigData, cloud computing and data mining. We also discuss current research problems in data science and provide concerns relating to a baseline of the data science industry.

1.1 Data Mining and Cloud Computing: The Prelude of BigData and Data Science

Since its creation, the Internet has infiltrated every aspect of a human's life. Every business has also adopted the Internet and World Wide Web for day-to-day operations. Data mining is the analysis of knowledge discovery over the Internet, in databases and/or in data storage such as data warehouses [12]. It is the process of analyzing data, searching for specific objects, or finding rules among data. For instance, data mining includes finding rules for customers who purchase one item and whether they will purchase another associated item. This is called the association rule, which is a major new development in data mining. Most data mining technologies are still statistical and artificial intelligence (AI) based approaches. From problem-solving point of view, anomaly detection, such as outlier or change findings, and association rule learning are two specific tasks in data mining in addition to the common tasks of pattern recognition. Pattern recognition usually means clustering, classification, modeling, and interpretation or summarization. Pattern recognition usually overlaps with artificial intelligence. Even it is very broad, AI mainly contains the research aspects related to searching, knowledge representation, reasoning, planning, learning, and decision making. Despite the differences in the names of terminologies, reasoning, planning, learning,

L.M. Chen (✉)
The University of the District of Columbia, Washington, DC, USA
e-mail: lchen@udc.edu

© Springer International Publishing Switzerland 2015
L.M. Chen et al., *Mathematical Problems in Data Science*,
DOI 10.1007/978-3-319-25127-1_1

3

and decision making will use pattern recognition technologies such as clustering, classification, and modeling. For instance learning a line or curve, we usually use regression technique that is a simple modeling method.

Cloud computing is the mechanism used in the organization of multiple computers and data sources that are randomly arranged in different locations with a variety of connections between them. It maximizes the flexibility of parallel computational power. Cloud computing provides computing related services to individual companies or users with multiple resources as utilities such as water, electricity, and phone services. Users do not have to build and maintain their own computing infrastructures such as databases, high performance computers, high speed Internet routers, and security systems.

Briefly, data mining on cloud computing can be viewed as the early stages of BigData processing. Data science is the new field of exploration that will be able to solve current and future problems related to BigData. BigData and data science provide many opportunities for scientists, engineers, and IT businesses. It also provides tremendous opportunities to mathematicians and computer scientists to discover new mathematics and algorithms. In this chapter, we will attempt to outline the different aspects of BigData and data science that may interest mathematicians or computer scientists. Data Mining and cloud computing have become precursors to BigData and data science.

In 2012, the United States federal government announced a so-called Big Data Research and Development Initiative. Its purpose is to improve "the ability to extract knowledge and insights from large and complex collections of digital data."

1.2   BigData Era

BigData technology is about data sets from many sources and collections of different formats of data. It also has the properties of massive storage and requires fast analysis through a large number of computing devices including cloud computers. It may yield revolutionary breakthroughs in science and industry. BigData is a phenomenon in current problems regarding data sets. The characteristics of BigData are: (1) large data volume, (2) use of cloud computing technology, (3) high level security, (4) potential business values, and (5) many different data sources. People used *3V* to describe the characteristics of BigData: **V**olume, **V**ariety, and **V**elocity [3].

Modern BigData computing is also called the petabyte age. A Petabyte (PB) is $1\,MB \times 1\,GB$. For instance, if Google gives each person $1\,GB$ for space storage and there are 1 billion people in the world who will use Google's service, the data volume would be $1\,G \times 1\,G = 1000\,PB$.

One software tool for BigData is called Apache Hadoop, which is an open-source software framework that supports data-intensive distributed applications and enables applications to work with thousands of computation-independent computers and petabytes of data. Hadoop was derived from Google's MapReduce and the

Google File System (GFS) [8, 11, 24]. MapReduce is a technological framework for processing parallelize-able problems across huge data sets using a large number of computers (nodes). In the meantime, MapReduce can take advantage of the locality of data, processing data near the storage in order to reduce distance transmission costs. A more sophisticated software system for BigData called *Spark* was developed by UC Berkeley. The Spark system is a Java programming based open source system for cloud computing. It makes data analytics fast than Hadoop. (https://amplab.cs.berkeley.edu/projects/spark-lightning-fast-cluster-computing/).

MapReduce consists of two major steps: "Map" and "Reduce." They are similar to the original *Fork* and *Join* operations in distributed systems, but they can consider a large number of computers that can be constructed based on the Internet cloud. In the Map-step, the master computer (a node) first divides the input into smaller sub-problems and then distributes them to worker computers (worker nodes). A worker node may also be a sub-master node to distribute the sub-problem into even smaller problems that will form a multi-level structure of a task tree. The worker node can solve the sub-problem and report the results back to its upper level master node. In the Reduce-step, the master node will collect the results from the worker nodes and then combine the answers in an output (solution) of the original problem.

Data science is a new terminology in BigData. How to make a Petabyte problem parallelize-able is the key to using Hadoop. What is Data Science? Data contains science. However, data science has a different approach than that of classical mathematics, which uses mathematical models to fit data and to extract information. In data science, some mathematical and statistical tools are required to find fundamental principles behind the data.

These fundamental principles include finding rules and properties of a data set, among different data sets, and the relationship of connectivity between data sets. The new research would be more likely to include partial and incomplete connectivity, which is also a hot topic in the current research of social networks. Previously developed technology such as numerical analysis, graph theory, uncertainty, and cellular automata will also play some role.

However, developing new mathematics is more likely the primary focus for the scientists. A good example is in face recognition and trying to find a person in a database with 10 million pictures when the pictures are randomly taken. Finding a best match for the new picture requires tremendous calculations.

It is related to the person's orientation in each picture. Let's assume that we have 100 computers but they are not available all the time. One can build a tree structures with these 100 computers. When a computer is available, it will get task from its father node. When it is not available, it will return its job to its father node.

Not every problem with a massive data set can easily be split into sub-problems. It depends on the connections of its graph representation. For instance, for an *NP*-hard problem, such as the traveling salesman problem, the sub-problem with less nodes does not help much in solving the whole problem. However, for a sorting problem, a solution of sub-problems would be helpful. If a merge-sort algorithm is used, Map-step can give a sub-problem to its worker nodes. The "reduce" step only requires linear time to merge them.

Table 1.1 Four paradigms of science [14, 17, 18]

Paradigm	Nature	Form of science	When
First	Experimental	Describing natural phenomena	Pre-Renaissance before 1600
Second	Theoretical	Modelling and generalization	Pre-computers before 1950
Third	Computational	Simulation of complex phenomena	Pre-BigData before 2000
Fourth	Exploratory, data-intensive	Statistical exploration, data mining	Now

Science and knowledge development is reliant on real-world data observations and collections. Kuhn and later Gray summarized four paradigms of science developments where BigData is at the fourth paradigm [14, 17, 18] (Table 1.1).

1.3 The Meaning of Data Sciences

The term data science was first coined in 2001 by Cleveland [7] as an extension of statistics focusing on data analysis. Data science has a close relationship to data mining in artificial intelligence and statistics, which was evidenced by the title of the first international journal on the topic published by Columbia University: *Journal of Data Science: an international journal devoted to applications of statistical methods at large.*

Data science was meant to consist of statistical applications for large data sets or extended data analysis methods kerneled by statistics [13]. This is still somewhat true. However, beginning in 2014, data science emerged into the spotlight in science and engineering. The reason was that BigData, initialed by industry for its ability to process data sets containing thousands of terabytes, required data science for extracting meaningful information from Internet based storage. Dhar [9] explained that the purpose of data science is to extract knowledge from data. It results in slightly better understanding of the data than from data mining since knowledge could be science, and therefore, not limited to the rules of the data set.

Dhar stated that "Data science might therefore imply a focus involving data and, by extension, statistics, or the systematic study of the organization, properties, and analysis of data and its role in inference, including our confidence in the inference."

What does extraction mean? Using John Tukey's quote, "The combination of some data and an aching desire for an answer does not ensure that a reasonable answer can be extracted from a given body of data." For example, one may have 1000 gigabytes of data but only a few kilobytes are useful to answer a question. Today, the key method for data science is still machine learning. A good example to explain how machine learning works is that for an online data analysis, when a discovery is found, we need to collect new samples and run the experiment to confirm or adjust the answer we get. This is called learning.

Why do we need a new term, data science, when we have statistics and data mining already? The size of the data sets should not be the only justification for requiring a new name. Dhar tried to answer this question with the following: "The short answer is data science is different from statistics and other existing disciplines in several important ways. To start, the raw material, the 'data' part of data science, is increasingly heterogeneous and unstructured-text, images, video-often emanating from networks with complex relationships between their entities." The data are not in the traditional formats. New data science must be capable to work on multiple types of data in integration, interpretation, and sense making. The software tools for data mining or data science may from difference discipline areas such as science, engineering, sociology, and even linguistics [9].

Another important distinction to statistics and traditional artificial intelligence is in the relationship among data sets or extracted data sets. This is related to integration, but not exactly the same. It can be used for integration, but when multiple data sets are extracted, rules and meanings are interpreted. The traditional method is to integrate immediately. However, an alternate method is to keep the data sets with their relations before integration. In such a way, different researchers can apply different integration technologies. It is important to note that relations are the foundation of integration.

Jeff Leek presented an opinion that states "the key word in 'Data Science' is not Data, it is Science." (http://simplystatistics.org) It is very true to some extent since any type of science relating to data is useful when the data are used to answer a question. However, without the changing of data size, there is no need to have a new term of data science. Traditional statistics may be sufficient to answer these questions.

We can summarize that *Data Science* is the science about the data especially about the massive data sets with veracity in their forms.

1.4 Problems Related to Data Science

To solve problems in the BigData era, we must consider using existing massive computational power such as cloud computing in many different ways. What can the new science discover in such a framework setting? What about other aspects including new mathematics that could be developed to fit the needs of the times. In this section, we list some of the problems and research aspects people are currently focusing on. Even though we do not have a very precise definition of data science, we could say that data science is about the study of:

1. The science of data;
2. Extracting knowledge from massive data;
3. Data and data set relations;

4. BigData processing; and
5. Visualization of massive data and human–computer interaction.

When dealing with extracting knowledge from data, "machine learning," the purpose is to discover patterns in data sets that have particular importance to data science. However the new problem is much harder when we consider a very large data set since there may only be a few kilobytes of data that are actually related to the model we are looking for among a few hundred gigabytes of data. The smart selection is not only sampling but also smart searching among data that is stored in different location, databases, or Internet.

The smart search is an online or simultaneous, synchronized and unsynchronized, immediate update of information when multiple machines are searching for an object or rule at the same time. This objective or target may also change. For instance, we have 10,000 machines running for such a search for a virus pattern, when one machine has determined the first component of the pattern, other machines will be assigned a new job for searching for other patterns. The correlation is the key for fast search results. What is the science about this particular problem? We use algorithmic for fast resolution and communication, statistical for sampling and testing, geometric to find the boundary of the pattern, algebraic for modification of equations, and analysis for comprehensive decision making. All of these play a part.

In a search, multiple machines and multiple tasks are running in a collaborative manner. Such a job has uncertainty based on where and when the first few tasks are completed. The simpler example is to search for a genetic disease in a group of people. We first need to search for a bad gene, a complete human genome will contain 4 gigabytes of data. When we need to find such a gene among 10 million people, this is now on the BigData scale. The 10 thousand machines will first begin to run their own Map-reduce. If a bad gene is found in one section of the gene profile, it will be announced to all other machines. All machines will then change their jobs to find the details and do an analysis for the total combined parts of the gene. Such a task is not one that existing technologies can solve automatically.

This is also related to when the machine is running for a long period of time. How do we change our target to record the current results and predictions? It is obvious that we need more sophisticated research to solve this problem. However, currently, the focus is on one scale and multiple machines in BigData.

Some general problems in Data Science and BigData include:

P 1.   Extracting knowledge and rules with the smart search.
P 2.   Finding the relationship and connectivity of incomplete data collections (searching by identifying the relations among BigData).
P 3.   Cloud data computing and topological data analysis.
P 4.   Algorithms for high dimensional data including manifold learning; fast algorithms for MapReduce, for protecting user from providers in Networking, security.
P 5.   Applications on Massive Data recovery.
P 6.   Advanced topics of data sciences in mathematics.

Another difficult problem relates to data science is that different sets of data are in different formats using different analysis models. After each calculations, how do we merge the results together as a single outcome?

We can use two examples to explain these: (a) At Walmart, grocery sales analysis can use decision trees to do predictions, but people can also purchase electronics in the same visit. We sometimes cannot use the same model to both sales. The final sale will then require a big sale portfolio function. (b) For data reconstruction, one part requires cubic spline and the other requires Gaussian normal distribution functions for the fitting. How do we merge them into a smooth function?

1.5 Mathematical Problems in Data Science

A data set is called a cloud if its data points are randomly arranged, but they are dense and usually have a lot of data points. In data processing, there are only two primarily data formats: random cloud data and array-based data. Random cloud data (also called scattered data) are stored in the vector form, and array-based data are images in 2D, 3D, or higher dimensional arrays.

Mathematically, let M be a cloud data set that is a set of data points in m-dimensional space or M could be the points in arrays. Data science should be able to solve the problems described below. Some of the problems were listed in [5].

Question 1. How do we find the orientation of the data set? We can use the principal component analysis to get the primary eigenvectors. We can continue to ask the problem related to BigData as follows: When data sets are stored in different databases or locations, how do we estimate the answer as we combine the answers from paretical results? We will discuss the methods in Chap. 2.

Question 2. How do we classify M into different classes? The best method is k-means if there is no time limit. We can also use neural networks and supporting vector machines to separate the data set. When M is stored in different computers with large volume, the efficiency and merge of results might yield some new problems. Machine learning is playing a major role in classification problems. We will discuss most important machine learning methods in Chap. 4. We also discuss them in Chaps. 2 and 5.

Question 3. How do we find the partition of the space using the data set? Even though the Voronoi diagram is the best choice, however, new algorithms that adapt to the big data sets need to put into a consideration.

Question 4. If some categorization of sample data is given, then in some instances we know the data classification of a subset of M. We can use k-nearest neighbor (k-NN) to find the classification for unknown data points. See Chaps. 2 and 4. Some problems will be related to Monte Carlo method in Chap. 7.

Question 5. How do we store the data points so that later we can retrieve the data points? We can use quadtree or R-tree to store the data and we can save a lot of space [5]. An introduction is given in Chaps. 3 and 5. This data format has particular importance in networking especially wireless data communication and computing. More research has been done recently.

Question 6.   If most of the data does not occupy the entire space, can we find a subspace to hold the data and not waste the remaining space? This problem is related to dimension reduction [15] and manifold learning [5, 22]. In this area, mathematicians can find great deals of new problems to do research with. We will explore the related problems in Chaps. 6 and 8 in this book.

Question 7.   How do we find the topological structure of the data sets? A technology called persistent homology analysis was proposed to solve this problem [5]. The topological structure of the data sets is now one of the major areas where mathematicians and computer scientists have put considerable attention. The geometric structure of massive data or BigData will be critical in data analysis. We predict there will be much more new findings in theory and practice. We will have Chap. 7 and part of Chap. 3 that are also related to this topic.

Question 8.   When the cloud data is dense enough to fill an entire space, we call the image M. We will use image processing methods to solve the related problem [19, 21, 23, 25]. In image processing especially today, video object search is a main concern. Videos generate big data. The search algorithm must be efficient for some surveillance cameras. The mean shift method is one of the most successful methods. We will discuss it in Chap. 5.

Question 9.   Image Segmentation is still fundamental in most of the image processing problems where classic and modern mathematics have contributed significantly in the past. However, the modern BigData requests more efficient and parallelized algorithms we still lack of. In Chap. 5, we will review popular image segmentation method as well as specialized techniques including variational methods and graph-cut methods. We will also discuss the cloud computing technology in application of image segmentation in Chap. 5. In Chap. 10, a detailed method combining machine learning and variational method will be presented.

Question 10.   For Smart Search we discussed in the above section, beyond the algorithm design and implementation in cloud computing environment. Some research in statistics can also be interesting. For instance, the smart search method with the consideration on tread-offs among time complexity, space-need, and accuracy of the results. In some occasions, one might must choose probabilistic method to avoid exponential time complexity. In this case, statistical and experiential studies may be necessary.

Question 11.   The NetFlix Problem is also called the matrix completion problem [4]. It is to find a way to fill all elements in a huge matrix $n \times N$ when some entries are known. This is also related to data recovery and reconstruction. A popular method to solve this problem is to assume a low-rank vector space then to generate the instance of the vectors in the matrix. The best result is at the level of $rN \log N$ [10, 20] where r is the rank of the (assumed) space and N is the number of columns of the big matrix. The NetFlix company wanted to find the missing data of a matrix M that records movie watching review. n is the number of different movies, and N represented the number of customers who watched the movie. Each element of the matrix could be just binary or the rating value given by a customer. We will discuss the related problem in Chap. 6. But will focus on geometric method. In a NetFlix competition, a team used the SVD method and Neural Networks won.

Question 12.   Subspace clustering is a special data segmentation problem [2, 15]. Mathematically, there is an equivalence between this problem and the NetFlix Problem (in high rank space) [10]. This is because that we can cluster the vectors into subspace that is in low-rank and use the method in Question 11. We can state this problem as follows: There are N samples (vectors). Can we find k-subspaces with (possibly different) low-dimensions to hold all data samples? A good survey is [22]. The existing method still uses matrix decomposition techniques such as SVD and PCA. With restricting the rank of each subspace to at most r, a good result is attained at $krNlog^2N$ for the requirement of elements in the matrix. These methods assumed that there is no noise. This is because a little noise will change the rank of the matrix in a significant way. When the rank is high, these methods will not take advantages.

This is also related to motion detection in video images we will discuss it in Chap. 5. An algorithm related to search method is discussed in Chap. 6 where the noise is allowed.

Subspace clustering is to learn the subspaces from samples in R^n. Then, assign each vector to its proper subspace. The components with small differences of two vectors should be treated as they have the same value on these components. So it is possible to do the reduction in dimensions for a subset of samples. If we can cluster all samples to subspace, then the union of the subspaces will be enough to represent the data sets. This is essentially a search problem. However, if we search every possible combination, the algorithm will be exponential.

Many problems in financial industry such as interest rate reconstruction and online learning for market data analysis are BigData related [1]. This book will discuss financial data analysis and reconstruction in Chap. 9.

1.6 Mathematics, Data Science, and Data Scientists in Industry

Mathematics studies numbers, shapes, and inferences. Data consists of numbers. A set of data has a shape, and the result is obtained by inferences. With such a prospective, data science is a division of mathematics. However, data science is not a theory. It is used for massive data processing. It is also used in industrialized mathematics with regard to fast algorithm design [6], data collection and storage, Internet communications, integration of intermediate results, or any combination of the above.

Many mathematicians will find jobs in the data science industry in the very near future. The following is a list of hints for young graduates on how to become a successful data scientist. If you have or would like to have any of the following traits on the list, it is most likely that you could potentially be a very good data scientist, even without a degree from the newly created advanced degree programs in data science.

These traits are:

1. Have knowledge in BigData. MapReduce, Hadoop, and Spark as well as other tools;
2. Be able to acquire data with different file formats like XML, CSV and SAS, SPSS;
3. Understand and be able to use image and video formats such as principle of image compression;
4. Be able to use tools and techniques for sampling and filtering data, along with data transformation; (Chap. 4)
5. Understand and use basic techniques of pattern recognition and data mining;
6. Understand and use basic techniques in statistical computing including some tools such as *SAS* or *R*;
7. Be able to analyze data using Hadoop or Spark;
8. Understand basic and some advanced techniques in machine learning techniques. Be able to use some tools such as *CAT* or Apache Mahout; and
9. Some hybrid learning technologies such as neural networks, genetic algorithms and fuzzy logic will be a plus.
10. Understand some basic algorithmic design technologies such as divide-and-conquer, greedy, and dynamical programming.
11. Understand computational methods such as curve fitting will be useful for special industry.

These are modified from http://www.thedevmasters.com/big-data-data-science/.

People who understand and use some of the above technologies will be able to find a job in data science industry. Some companies have already started to have data science training lessons. For example, http://www.thedevmasters.com/big-data-data-science/. Some companies have service for image-based recognition and object tracking.

The general purpose computing system (the service providers) for cloud computing and BigData are mainly three: (1) Amazon's AWS system, (2) Google's Cloud, and (3) Microsoft's Azure.

1.7   Remark: Discussion on the Future Problems in Data Science

If data science is about the science of the data and not only about the data itself, the relationship among data or data sets must be a factor. Finding the connection of data and data sets is also the primary concern of data science. Science is about knowledge, so data science is about the knowledge of data (extraction) among data points or among the subsets of data collections (data relations).

The best example of data science is still the Google search page ranking (Mike Loukides), where "Google's breakthrough was realizing that a search engine could use input other than the text on the page. Google's PageRank algorithm was among the first to use data outside of the page itself, in particular, the number of links pointing to a page. Tracking links made Google searches much more useful, and PageRank has been a key ingredient to the company's success."

What is data science? Mike Loukides said : "The future belongs to the companies and people that turn data into products."

Data Science is a composite of a number of pre-existing disciplines. It is a young profession and academic discipline. The term was first coined in 2001. Its popularity has exploded since 2010, pushed by the need for teams of people to analyze the big data that corporations and governments are collecting. The Google search engine is a classic example of the power of data science. Other online algorithms can be found in [16, 21].

Data science is a discipline that incorporates varying degrees of data engineering, scientific method, math, statistics, advanced computing, visualization, hacker mindset, and domain expertise. A practitioner of Data Science is called a Data Scientist. Data Scientists solve complex data analysis problems.

In this book, we mainly introduce the existing methods for data science and BigData. We also list multiple problems that still challenge us today. Finally, we point out some of the future problems with this topic.

This book will focus on the problems in data science. That is especially written for mathematicians from computer scientists' point of view.

In the satellite conference on Mathematical Foundation of Modern Data Sciences of ICM 2014 (International Congress of Mathematicians, Seoul, Korea, 2014), we organized a Panel discussion on the problems where mathematicians could play the central role. Now, we list some of the problems suggested and discussed:

S 1   Can one use numerical algebra and PDE to be the edge descriptor of objects in image segmentation? What is the relationship to the existing level-set method? This problem is suggested by Bo Yu at Dalian Tech Univ.

S 2   Tian Liu at Peking University raised the issue of computing models. What is a better computational model for BigData and what differs it from traditional model? Li Chen added that Robert Soare at University of Chicago said that Internet computing is the Oracle Turing Machine, should we agree?

S 3 Bo Jiang at Dalian Maritime University pointed out if we compute the problem of the shortest path when we have three million data points. How do we perform the algorithm? This problem may need to do analysis for the convex Hull problem in high dimension. Reinhard Klette added that 3D convex Hull is critical. Binhai Zhu added the topological relationship of data points including Voronoi diagrams in case we have massive data points. How do we solve the problem using multiple computers or cloud computing?

S 4 XiaoDong Liu would like to see a theory that establishes a theory in mathematics for human learning & understanding for data science, current fuzzy set theory can be improved.

S 5 Reinhard Klette suggested to establish a unified theory for randomness, theory, and practices not only about the detail or particular technology, but also for unification the three categories. How will cloud computing be taken part in this unification?

S 6 Jijun Liu liked an idea that says "BigData is the inversion problem (Wienan E)." Can we use 1000 computers to solve a problem in PDE inversion with much better accuracy? He mentioned regularization theory for solving problems with lack of data and instability. (Refer to the book entitled, "The Mathematical Sciences in 2025.")

S 7 Zhixun Su added a question: What is the regular part of the data? Can we learn the regularization rules? Other discussions include: Can we combine geometry part method and statistical method? Can we mixed statistics with geometry for manifold learning optimization from different way such as manifold learning for local metric?

S 8 Lizhuang Ma and Bo Yu would consider QQ.com users and the sparse matrix for very large number of users. Can we use spline, multi-spline, or NURBS to solve these problems? Other questions such as patterns of data and similarity of data internal connection were discussed when using these techniques.

S 9 Renhong Wang observed a phenomena: Can we develop a technology that adds a dimension, so we can easily get any secrete data? For instance, a circle in 2D is secure in holding data inside of the circle, but if we add a dimension, then we can easily get the thing from the circle and its internal part.

References

1. J. Abernethy, Y. Chen, J.W. Vaughan, Efficient market making via convex optimization, and a connection to online learning. ACM Trans. Econ. Comput. **1**(2), Article 12 (2013)
2. L. Balzano, R. Nowak, A. Szlam, B. Recht, k-Subspaces with missing data. University of Wisconsin, Madison, Technical Report ECE-11-02, February 2011
3. R.E. Bryant, R.H. Katz, E.D. Lazowska, Big-Data computing: Creating revolutionary breakthroughs in commerce, science, and society (2008). Computing Research Consortium, at http://www.cra.org/ccc/resources/ccc-led-white-papers/. Accessed 20 September 2013

4. E.J. Candes, T. Tao, The power of convex relaxation: near-optimal matrix completion. IEEE Trans. Inf. Theory **56**, 2053–2080 (2010)
5. L.M. Chen, Digital and Discrete Geometry: Theory and Algorithms, Springer, 2014
6. L. Chen, How to be a good programmer. ACM SigAct News, **42**(2), 77–81 (2011)
7. W.S. Cleveland, Data science: an action plan for expanding the technical areas of the field of statistics. Int. Stat. Rev./Revue Internationale de Statistique **69**(1), 21–26 (2001)
8. J. Dean, S. Ghemawat, MapReduce: simplified data processing on large clusters. Commun. ACM **51**(1), 107–113 (2008)
9. V. Dhar, Data science and prediction. Commun. ACM **56**, 12–64 (2013)
10. B. Eriksson, L. Balzano, R. Nowak, High rank matrix completion, in *Proceedings of International Conference on Artificial Intelligence and Statistics* (2012). http://jmlr.csail.mit.edu/proceedings/papers/v22/eriksson12/eriksson12.pdf
11. Hadoop: Open source implementation of MapReduce. http://lucene.apache.org/hadoop/ (2014)
12. J. Han, M. Kamber, *Data Mining: Concepts and Techniques* (Morgan Kaufmann, Los Altos, CA, 2001)
13. T. Hastie, R. Tibshirani, J. Friedman, *The Elements of Statistical Learning: Data Mining, Inference, and Prediction*, 2nd edn. (Springer, New York, 2009)
14. T. Hey, S. Tansley, K. Tolle, Jim Grey on eScience: a transformed scientific method, in *The Fourth Paradigm: Data-Intensive Scientific Discovery. Redmond: Microsoft Research*, ed. by T. Hey, S. Tansley, K. Tolle (2009), pp. xvii–xxxi, Microsoft Research
15. K. Kanatani, Motion segmentation by subspace separation and model selection, in *Proceedings. Eighth IEEE International Conference on Computer Vision, 2001. ICCV 2001*, vol. 2 (2001), pp. 586–591
16. R. Kennedy, B. Laura S.J. Wright, C.J. Taylor, Online algorithms for factorization-based structure from motion. ArXiv e-print 1309.6964 (2013). http://arxiv.org/abs/1309.6964. 2014 IEEE Winter Conference on Date of Conference on Applications of Computer Vision (WACV) 24 26, 37 44 (2014)
17. R. Kitchin, Big Data, new epistemologies and paradigm shifts. Big Data Soc. **1**(1), 1–12 (2014)
18. T. Kuhn, *The Structure of Scientific Revolutions* (University of Chicago Press, Chicago, 1962)
19. J.-C. Pinoli, *Mathematical Foundations of Image Processing and Analysis*, vols. 1, 2 (Wiley, New York, 2014)
20. B. Recht, A simpler approach to matrix completion. J. Mach. Learn. Res. **12**, 3413–3430 (2011). arXiv:0910.0651v2
21. A. Teran, Real-time multi-target tracking: a study on color-texture covariance matrices and descriptor/operator switching, University of Paris Sud, Paris XI, Ph.D. Thesis, 2013
22. R. Vidal, A tutorial on subspace clustering. Johns Hopkins Technical Report, 2010. http://www.cis.jhu.edu/~rvidal/publications/SPM-Tutorial-Final.pdf
23. R. Vidal, R. Tron, R. Hartley, Multiframe motion segmentation with missing data using PowerFactorization and GPCA. Int. J. Comput. Vis. **79**(1), 85–105 (2008)
24. T. White, *Hadoop: The Definitive Guide*, 4th edn. (O'Reilly, Sebastopol, CA, 2015)
25. A. Yang, J. Wright, Y. Ma, S. Sastry, Unsupervised segmentation of natural images via lossy data compression. Comput. Vis. Image Underst. **110**(2), 212–225 (2008)

Chapter 2
Overview of Basic Methods for Data Science

Li M. Chen

Abstract Data science utilizes all mathematics and computer sciences. In this chapter, we give a brief review of the most fundamental concepts in data science: graph search algorithms, statistical methods especially principal component analysis (PCA), algorithms and data structures, and data mining and pattern recognition. This chapter will provide an overview for machine learning in relation to other mathematical tools. We will first introduce graphs and graph algorithms, which will be used as the foundation of a branch of artificial intelligence called search. The other three branches of artificial intelligence are learning, planning, and knowledge representation. Classification, also related to machine learning, is at the center of pattern recognition, which we will discuss in Chap. 4. Statistical methods especially PCA and regression will also be discussed. Finally, we introduce concepts of data structures and algorithm design including online search and matching.

2.1 "Hardware" and "Software" of Data Science

Data storage, data communication, security, and computing machinery can be considered as the "hardware" in data science. Artificial intelligence (AI), statistical methods, algorithm design, and possible new mathematical methodologies can be viewed as the "software" of data science.

Existing methods play a primary role in data science today. There are three aspects that are important to modern data related problems. We want to give brief statements on each of them. For mathematicians, these are part of the motivation for why we are interested in data science problems and how we start to tackle these problems. For some researchers, the motivation is figuring out what we should do to find the new method to solve a challenging problem.

L.M. Chen (✉)
The University of the District of Columbia, Washington, DC, USA
e-mail: lchen@udc.edu

© Springer International Publishing Switzerland 2015   17
L.M. Chen et al., *Mathematical Problems in Data Science*,
DOI 10.1007/978-3-319-25127-1_2

2.1.1 Searching and Optimization

Search is one of the most important tasks in AI [34]. General search, in fact, is not a topic of AI, but it belongs to algorithm design and usually relates to graph search. For instance, finding a special object in a graph or finding the best path. When the area in which we are searching is very large, this type of algorithm is not feasible, meaning that we can not complete the search in an allotted time frame. AI search will look for a best or near best way to limit some of the unnecessary paths that unlikely contains the object or target we are looking for.

For instance, A^* search is a type of search based on the so-called breadth first search method in graph theory, which finds the best possible direction to look for the answer.

2.1.2 Decision Making

Everyday, we need to make many decisions. What time should I leave the house for work? What I am going to eat for lunch? Can I come home earlier to avoid traffic? Decision making usually refers to statistical methods and discrete (algorithmic) methods.

Everyone knows the regression method to find a line that best fits sample data in 2D. We usually use the least square method to find such a line, which is a statistical method. Another method is called neural networks that employs the least square method too. It uses multiple layers where each layer performs linear transformation like operations. The process also adds some nonlinear procedures and automated adjustments of parameters based on the least squares.

The decision tree method is a typical algorithmic method [12]. It does not look at the decision for the whole data set. It makes a decision at each individual step. For instance, John has a meeting in the morning, so he leaves home 30 min earlier than usual. We will set up a rule such as "if someone has a meeting in the morning, then he or she needs to leave home 30 min early." When John meets heavy traffic, he will choose to use a toll-road. We then have another if-then rule.

The decision tree method preforms a set of "IF-THEN" statements. The decision is made based on each condition that has occurred. The structure of the method or algorithm is like a binary tree where each node will test if the condition is met. Then, the algorithm will direct John to go left or right in the child node.

2.1.3 Classification

Classification is the placement of certain samples into categories [34, 39]. If the categories are predefined, then we call it supervised classification. Otherwise, we call it unsupervised classification.

The problem of classification is everywhere. For instance, to measure if a person is overweight, we use something called BMI, which is a function of weight and height. When $f(W, H) = (W/H^2) \times 703$, where W = (Weight in Pounds), H = (Height in inches). If the value is above the average, we say the person is overweight. Otherwise, the person is normal. Using this method, people are categorized into two classes: (1) normal and (2) overweight. If we are only interested in people who has the same hight, say 72 inches. So the formula will be $f(W) = (W/(72 \times 72)) \times 703$. The new function that determines the classification is linear. Therefore we call it the linear classifier. Assume the clip-level for the normal class is under 25, so everyone who has BMI less than 25 is healthy (with height = 72 inches). This is also called a supervised classification (since we have predefined a clip-level value).

In another example, we have a picture that we want to separate into different objects. However, we do not know how many objects are in the picture. We can only use the intensity of the image to do the classification. The pixels (picture elements) similar in color or gray scale will be treated as the same object if they are close to each other. This classification is an unsupervised classification.

2.1.4 Learning

For some problems, it is hard to generate rules before we start to process the data such as a function. We generate the model dynamically, meaning that we need to use the samples to obtain the parameters of a classification system. This method usually refers to a learning method. Let us assume that we have satellite data sets from a GPS system. We want to classify the geographical areas into cities, forests, and water. The volume of data is so big that we ask some experts to select 1000 points (locations) and then identify those points as city, forest, or water. Then, we recollect the GPS image data from the database to retrieve the values of the image intensity (with different bandwidths so they have a total of seven bands for seven pictures). We now have data for these 1000 points $< F, v >$. Note that the decision about these 1000 data points is made by experts. We also have the location of the data as well. So we have $(x, y, f_1, ..., f_7, v)$ where v is a value for city, forest, or water. We call this data set the samples S. In order to make a learning system, we split S into two sets: the training set S_{tr} and the testing set S_{te}. We usually request $S_{tr} \cap S_{te} = \emptyset$. However, in statistical learning methods, the researcher usually intends to make S_{tr} and S_{te} overlap for more profound analysis.

We use the training set to find the model for the data and then use the testing set to test the model system we used to see if it is correct. Using the simple example of linear system classification, we can use S_{tr} to find a linear function for finding the PMI of a city population. (This is not a global solution, so we decided to find the parameter or coefficients of the linear equation, such as c1 and c2.) Then, we want to see if this linear model works or not. We can use S_{te} to test and find the correctness of the testing data samples.

We could also randomly reselect S_{tr} and S_{te} to test this set. If we are satisfied, we can use this model. If not, we can select another method. This way is called machine learning [34].

In this chapter, we review basic methods for data science.

2.2 Graph-Theoretic Methods

The graph in mathematics is a general structure for discrete objects. Theoretically, every computerized data storage or data processing mechanism can be deduced into an operational process of graphs. A graph G consists of two sets V and E, where V is the set of vertices and E is the set of pairs of vertices called edges. An edge is said to be incident to the vertices if it joins [3, 12, 17].

2.2.1 Review of Graphs

Graph $G = (V, E)$ is called a simple graph if every pair of vertices has at most one edge that is incident to these two vertices and there is no loop $(a, a) \in E$ for any $a \in V$.

In this book, assume that $G = (V, E)$ is an undirected graph, which means that if $(a, b) \in E$, then $(b, a) \in E$ or $(a, b) = (b, a)$. a and b are also called ends, endpoints, or end vertices of edge (a, b). It is possible for a vertex in a graph to not belong to any edge. V and E are usually finite, and the order of a graph is $|V|$, which is the number of vertices. The size of a graph is linear to $\max\{|V| + |E|\}$, meaning that it requires this much memory to store the graph.

The degree of a vertex is the number of edges that incident with (or link to) it. A loop is an edge that links to the same vertex. A weighted graph means that we assign a number on each edge that may indicate the distance between the two vertices.

If (p, q) is in E, then p is said to be adjacent to q. Let $p_0, p_1, \ldots, p_{n-1}, p_n$ be $n + 1$ vertices in V. If (p_{i-1}, p_i) is in E for all $i = 1, \ldots, n$, then $\{p_0, p_1, \ldots, p_{n-1}, p_n\}$ is called a path. If $p_0, p_1, \ldots, p_{n-1}, p_n$ are distinct vertices, then the path is called a simple path.

A simple path $\{p_0, p_1, \ldots, p_{n-1}, p_n\}$ is closed if (p_0, p_n) is an edge in E. A closed path is also called a cycle. Two vertices p and q are connected if there is a path $\{p_0, p_1, \ldots, p_{n-1}, p_n\}$ where $p_0 = p$ and $p_n = q$. G is also called connected if every pair of vertices in G is connected. In this book, it is always assumed that G is connected (unless otherwise specified).

There are three algorithms that are important to data science mentioned in this book [12]. These algorithms are: (1) The breadth first search algorithm, (2) The shortest path algorithm, and (3) The minimum spanning tree algorithm.

The maximum flow and minimum cut algorithms are also related to this topic. The data structures that refer to these two techniques are queues and stacks made by adjacency lists.

2.2.2 Breadth First Search and Depth First Search

Breadth first search is a fast search approach to get a connected component on a graph. It begins at a vertex p and searches for all adjacent vertices. Then, it "inserts" all adjacent points (neighbors) into a queue. "Removing" a vertex from the queue, the algorithm calls the point p and then repeatedly finds p's neighbors until the queue is empty. Marking all the vertices we visited, the marked vertices form a connected component. This technique was introduced in [12].

Algorithm 2.1. Breadth first search technique for all point-connected components

Step 1: Let p_0 be a node in G. Set
$$L(p_0) \leftarrow * \text{ and } QUEUE \leftarrow QUEUE \cup p_0$$
i.e., labeling p_0 and p_0 is sent a queue $QUEUE$.

Step 2: If QUEUE is empty, go to Step 4; otherwise,
$$p_0 \leftarrow QUEUE \text{ (top of } QUEUE\text{). Then,}$$
$$L(p_0) \leftarrow 0.$$

Step 3: For each p with an edge linking to p_0,
do
$$QUEUE \leftarrow QUEUE \cup p \text{ and } L(p) \leftarrow *. \text{ Then, go to Step 2.}$$

Step 4: $S = \{p : L(p) = 0\}$ is a connected part.

Step 5: If p is an un-visited vertex, then let $p_0 = p$. Repeat Step 1.
Otherwise Stop.

Breadth first search only visits a new node twice: once when the node is inserted into the queue and once when the node is removed from the queue. Therefore, this algorithm is a linear time algorithm and uses a fast search approach. Depth first search is similar to breadth first search. Depth first search will find new nodes continuously until no more new nodes can be found. When the algorithm runs through a complete iteration, it saves all visited nodes in a stack. Since we only go with a path to try and find the "deepest" node, there are other branches we may have missed in the first try. Therefore, the algorithm returns to each visited node in the order it was pushed in stack (first in, last out). We check other branches by popping out nodes. After all nodes are popped out, this algorithm will have found all the nodes in a component. This algorithm will visit all edges at most twice. Therefore, it is also a linear algorithm on the edges.

2.2.3   Dijkstra's Algorithm for the Shortest Path

Finding the shortest path in the weighted graphs is a popular topic in graph theory. Two of the most important algorithms are Dijkstra's algorithm and the Bellman–Ford algorithm [12]. Dijkstra's algorithm is faster but cannot take negative edges.

The idea of the shortest path algorithm consists of the following. Beginning at the vertex, this is the departing node. Then, we record the distance from the departing node to all of its neighbors. The record may not be the shortest path involving these notes, but we update the record and update the values on the nodes, a process called relaxation. After that, we extend the notes by one more edge, do a relaxation, and continue to repeat these two steps until we have reached every node in the graph.

The algorithmic technique used in this problem is called dynamic programming. Even though there may be an exponential number of paths from one vertex to another, if we are only interested in the shortest path, we only need to care about the length we travel. We do not have to pass a certain vertex. There are only $n(n-1)/2$ pairs of vertices. If we update the minimum distance while we calculate, this is called dynamic programming.

Algorithm 2.2. Dijkstra's algorithm for the shortest path
  Step 1: Let $T = V$. Choose the source point a
         $L(a) = 0; L(x) = \infty$ for all $x \in T - \{a\}$.
  Step 2: Find all neighbors v of a node u with the value of L at $L(v) = L(u) + w(u, v)$.
         $T \leftarrow T - \{v\}$
  Step 3: For each x adjacent to v, do
         $L(x) = max\{L(x), min\{L(v), L(x) + w(x, v)\}\}$
  Step 4: Repeat steps 2–3 until T is empty.

2.2.4   Minimum Spanning Tree

The minimum spanning tree is to find a tree in a connected weighted graph G such that the tree contains all vertices of G and the total weights are at the minimum.

Kruskal's algorithm can be used to find the minimum spanning tree. In this method, a technique called the greedy algorithm is used. The methodology is to pick up the best one (the shortest edge currently available) in each stage to find an ultimate solution.

In Kruskal's algorithm, the tree T initially contains all vertices but no edges. Because of this, it starts as an iterative process, adding an edge to T under the condition that the edge has the minimum weight. We continue to add a minimum edge to T as long as it does not allow for a cycle to appear in T. When T has $|G| - 1$ edges, the process stops.

Algorithm 2.3. Kruskal's Algorithm for finding a minimum spanning tree T

Step 1   Sort the edges based on the weights of the edges from smallest to largest.
Step 2   Set initial tree T to be empty.
Step 3   Select an edge from the sorted edge list and add it to T if such an added
          edge does not generate a cycle.
Step 4 T would be the minimal spanning tree if $|V| - 1$ edges are added to T.

The principle where we always make a minimum or maximum selection is called the greedy method. In artificial intelligence, the greedy method is constantly applied in applications [34]. In most cases, the method may not obtain an optimal solution but rather an approximation. However, for the minimum spanning tree, the greedy method would reach the optimal solution.

2.3   Statistical Methods

For a set of sample data points in 2D, the method of principal component analysis (PCA) can be used to find the major direction of the data. This is essential to data processing. The method is also related to regression analysis and other methods in numerical analysis and statistical computing [9, 24, 31]. Even though there are many ways to explain PCA, the best one is through statistics.

For a set of sample points, $x_1, x_2 \cdots, x_n$, the mean is the average value of the x_is

$$\bar{x} = \frac{1}{n} \sum_{i=1}^{n} x_i.$$

We may use a random variable X to represent this data set.

Variance is the measure of how data points differ from each other in a whole. The sample variance of the data set or the random variable X is defined as

$$Var(X) = \frac{1}{n} \sum_{i=1}^{n} (x_i - \bar{x})^2.$$

The square root of the variance is called the standard deviation σ_X, i.e. $Var(X) = \sigma_X^2$. Note that people sometimes use $n - 1$ instead of n in the formula for an unbiased estimate of the population variance.

For simplicity, let $X = \{x_1, x_2 \cdots, x_n\}$ and $Y = \{x_1, x_2 \cdots, x_n\}$. The covariance of X and Y is defined as

$$Cov(X, Y) = \frac{1}{n} \sum_{i=1}^{n} ((x_i - \bar{x})(y_i - \bar{y})).$$

We also can define the correlation of X and Y as follows:

$$Cor(X, Y) = \frac{Cov(X, Y)}{\sigma_X \cdot \sigma_Y}.$$

Now, let's introduce the PCA, an essential technique in data science. Given a set of 2D points, $(x_1, y_1), \cdots, (x_n, y_n)$, we can treat X as a random variable of the first component and Y as the second component of the 2D vector. We define the covariance matrix as the following.

The geometric direction of a set of vector data points can be determined through PCA. The largest eigenvalue of the matrix indicates the major axis by calculating the corresponding eigenvector.

$$M(X, Y) = \begin{pmatrix} Cov(X, X) & Cov(X, Y) \\ Cov(X, Y) & Cov(Y, Y) \end{pmatrix} \tag{2.1}$$

PCA can be applied to find the major axis of the data in 2D. See Fig. 2.1.

In general, this method can be extended to analyze data in m dimensional space. Assume we have m random variables in X_i and n samples for each variable. In other words, we have N sample points $p_i = (x_1, x_2, \cdots, x_m)$, $i = 1, \ldots, N$. We can extend the covariance matrix to be $m \times m$. Therefore, we will have m eigenvalues, $\lambda_1, \ldots, \lambda_m$. We assume that $\lambda_1 \leq, \ldots, \leq \lambda_m$. The eigenvector V_i of λ_i indicates the i-th principal component. This means that most of the sample data is along the side of vector V_i, compared to vector V_j, if $j > i$.

For data storage purposes, we can determine that most of the data is covered by the first few principal components. We can use linear transformation to attach this to the original data and store the transformed data in a much lower dimension to save space. This is why PCA is one of the most effective methods in many applications in BigData and data science today.

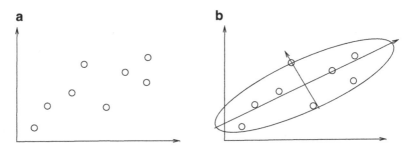

Fig. 2.1 Principal component analysis: (**a**) original data points, and (**b**) eigenvectors of the covariance matrix

2.4 Classification, Clustering, and Pattern Recognition

Classification usually refers to the supervised classification. In this method, we want to know the categories of data points. We usually collect some samples whose classifications are known. It is also called pattern recognition meaning that we want to put a new sample into its pattern category.

Clustering is called unsupervised classification. It is a process where we do not know the pattern. This means that we can only partition the data into several categories [28, 34, 39].

In this section, we introduce the most popular classification methods: k-Nearest Neighbor Method (kNN), and k-Means Method. The methods discussed in this section will be used in image data analysis in Chap. 5 [23, 26, 27, 32, 37].

2.4.1 k-Nearest Neighbor Method

The kNN method is a supervised classification method. It is often used in pattern recognition.

Let us take a sample subset of data points $S = \{x_1, \cdots, x_n\}$. For each x_i we know its classification. If x is a new data point, the shortest distance $d(x, x_i)$, $i = 1, \ldots, n$, indicates the class containing x_i that should include x. This is also called the nearest neighbor method.

kNN is a generalization of the nearest neighbor method. We find the k nearest neighbors in S for x of x in S, then we determine that x will be contained in the class that also contains most of those neighbors [22, 34, 39].

2.4.2 k-Means Method

The k-means method is an unsupervised classification method. Given a set of vectors, if we want to classify the data set into two categories, we can find two centers such that the total summation of the distances of each data point to its center will be minimum.

In Fig. 2.2, we first select two initial vectors as the "centers"(called sites) of the two categories. Second, for each element P in the set, we put it into one of the categories and record the distance from the center. Third, we update the two center locations such that the total summation of the distance becomes smaller. The algorithm will halt if there is no improvement when we move the two centers.

The k-means algorithm can be explained below: Based on the initial k sites, partition the space using the Voronoi diagram. Then, we can move the center to the geometric centroid of the new partition before we recalculate the results. This is an iterated process. The only problem is with the local minimum, meaning that this algorithm may converge to the local minimum for best result. In other words, the algorithm will stop after a local minimum is attained [39].

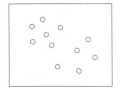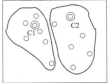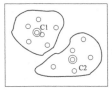

Fig. 2.2 The k mean algorithm

We present the k-means algorithm below:

Algorithm 2.4. The k-means algorithm

Input: For a set of data points $X = \{x_1, \cdots, x_n\}$, each x_i is an m-dimensional real vector.

Output: k-means tries to find $P = \{c_1, c_2, \cdots, c_k\}$, where each c_i is a vector. Then, all the data points will be partitioned into k subsets of X, where S_i associated with c_i and satisfies the following:

$$\min \sum_{i=1}^{k} \sum_{x_j \in S_i} \left\| \mathbf{x}_j - c_i \right\|^2. \tag{2.2}$$

Step 1   Randomly select k vectors $c_1, c_2, \cdots, c_k$ as the initial centers.

Step 2   For each data point x_i, calculate the distance $d(x_i, c_j)$, setting x_i as S_t if $d(x_i, c_t)$ is the smallest.

Step 3   Calculate the new geometric center of S_j by

$$c_j = \frac{1}{|S_j|} \sum_{x_i \in S_i} x_i$$

Step 4   Repeat Step 2 until the total square of the distances in (2.1) can no longer be improved.

2.5 Numerical Methods and Data Reconstruction in Science and Engineering

The most popular problem in data reconstruction is fitting a curve or a surface such as Bezier polynomials and B-Splines. Data fitting has two meanings: interpolation and approximation. Interpolation means fitting the data exactly on the sample values while approximation means setting a fitted function near the sample values. Other methods include data reconstruction from projection, such as CT scans in medical imaging, and discrete fitting, which can be found in [8, 20].

The most useful feature in data reconstruction is finding the missing data elements of massive data collections in data science. IT is especially important to fill the gaps in financial marketing data. For instance, during holidays we do not have actual New York stock exchange information. However, the industry is not on vacation. Therefore, the data information could be restored by filling in the predicted information. This technique has been used for a long time. See Chap. 9.

In this section, we will not do a comprehensive review of numerical and computational methods for data fitting. We will only discuss a principle of data fitting using the least squares method. For more information on the numerical method, we recommend the book Numerical Recipes and [8, 31].

Assume we have n sample data points in (x_i, y_i). These points are on a straight line, but the sampling process may have some observation errors. Based on the equation of a line (2.2), we have:

$$y_i = a + bx_i , i = 1, \cdots , n \qquad (2.3)$$

However, since there are observation or sampling errors, this equation may not always hold true for all sample points. In other words, $e_i = y_i - (a + bx_i)$ is not always zero. What we can do is find a and b such that the summation of the square of errors $((e_i)^2)$ is minimized. So, it is called the least squares method .

$$\text{Minimize } E = \Sigma_{i=1}^{n}(e_i)^2 \qquad (2.4)$$

where E is the function of a and b so that $E = E(a, b)$. To get the minimum $E(a, b)$, according to the extreme value theorem in calculus, the following two equations must hold:

$$\frac{\partial E}{\partial a} = \frac{\partial}{\partial a}(\Sigma(y_i - (a + bx_i))^2) = \frac{\partial}{\partial a}(\Sigma - 2(y_i - (a + bx_i)))$$

$$\frac{\partial E}{\partial b} = \frac{\partial}{\partial b}(\Sigma(y_i - (a + bx_i))^2) = \frac{\partial}{\partial b}(\Sigma - 2x_i(y_i - (a + bx_i))).$$

Thus, we have a system of linear equations:

$$na + (\Sigma x_i) \cdot b = \Sigma y_i$$

$$(\Sigma x_i) \cdot a + (\Sigma x_i^2) \cdot b = \Sigma(x_i \cdot y_i).$$

We can easily solve these equations. In Chap. 9, we give some detailed descriptions of curve fitting and its application to financial industry.

2.6   Algorithm Design and Computational Complexity

Algorithms are at the heart of computer science [12]. Everything related to data processing can be viewed as a type of algorithm design. However, in some cases, such as linear regression, the mathematical part of the algorithm dominates the algorithm. Therefore, we do not really think of it as a typical algorithm, which means that the procedure has some complexity and is not straightforward.

The Euclidean algorithm that is used to find the greatest common divisor (GCD) is referred to as the first algorithm since the procedure is more important in yielding the results than the mathematical formula derivation.

Many mathematical algorithms for data processing do not care for data structure or the format of the data stored in computer memory or computer disks. This is because we used to focus on the mathematical part of algorithm design. However, in data science, as we have explained, data is stored in different forms and different places. Data structure becomes the key to efficient processing.

2.6.1   Concepts of Data Structures and Databases

Data structures are fundamental to data science. It is the way different data types are held in storage or memory for fast processing. The simplest data structure is the array. We will first introduce the concept of the queue, stack, linked list, and tree. Then, we will focus on complex data structures such as quadtrees, octrees, and R-trees. Those data structures are often used in geometric data storage and massive data collections.

Databases are special systems that hold a large amount of data. A database system such as Oracle or MySQL contains functions that can be called to store and obtain data samples in the system. A data structure will be implemented in a database system. For instance, Oracle is a relational database. It means that a table or an array is the basic data structure.

A record is a row of data in the array (or the table) that stores all information about an individual. For instance, we store all information about a student at a university in a record. The record contains the student's name, ID number, email address, etc. A column of the array (or the table) stores all information about a property of the data set, such as the student ID number for all students at the university.

A database system not only contains data, but also includes functions that can retrieve the data. For example, if we only need 100 students midterm test scores, we can get this data by asking the system. This is called a query. This query is based on the relation of the table. (A relation in mathematics is an algebraic structure that is a collection of select vectors.) Data science must be able to deal with different types of database systems.

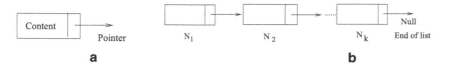

Fig. 2.3 A linked list: (**a**) a node configuration, (**b**) an example of a linked list

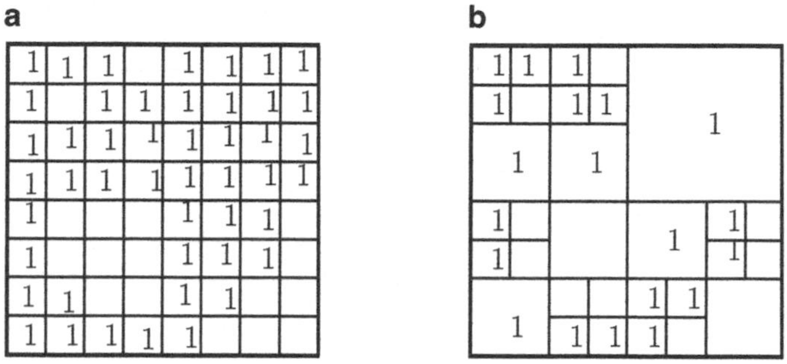

Fig. 2.4 An example of Quad-trees: (**a**) the original image, and (**b**) the quad-tree representation of (a). [7, 9]

2.6.2 Queues, Stacks, and Linked Lists

Queues, Stacks, and Linked Lists are efficient data structures to data science. To understand these data structures is essential to process massive data. A queue is a special data arrangement where the first entry is removed first; a stack performs differently in that the last data entry is removed first. A linked list is a collection of nodes that contains data and a pointer to the next node. See Fig. 2.3.

A tree is similar to a linked list. A node in a tree may contain two or more pointers. A binary tree is a tree where each node only has two pointers.

2.6.3 Quadtrees, Octrees, and R-trees

Quadtrees, Octrees, and R-trees are usually used for geometric data storage and retrieval.

The quadtree is for 2D data sets or objects. The root of a quadtree contains a 4-subtree. If we partition a 2D square into 4 small, identical subsquares, then each subtree will represent a subsquare. When storing a binary image, we can save space when a subsquare contains no information (it is blank) [20, 30, 33].

The quadtree method partitions a 2D space into four equal quadrants, subquadrants, and so on. We stop at a node called a leaf if all the elements represented by the node have the same value. We continue decomposing if this is not the case. See Fig. 2.4 [9].

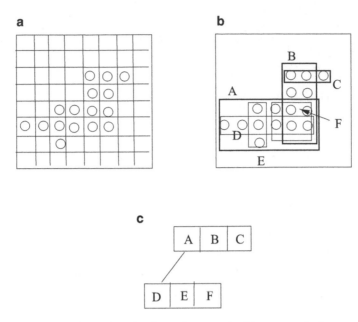

Fig. 2.5 *R*-Trees: (**a**) original data, (**b**) data points contained in rectangular boxes, and (**c**) tree hierarchy of boxes

The depth or height of the tree is at most $\log_2 n$ if we are dealing with an $n \times n$ region. We can save space since one leaf may represent many pixels in real images.

The octree uses the same principle as the quadtree, but it is used for 3D data storage. This is because we can split a 3D cube into 8 subcubes. Therefore, octrees are used for spatial data representation [36].

Each node in an octree represents a cubic region. We subdivide it into eight octants that will be represented by eight subnodes. When the values of the subregion are the same. The corresponding node will be a leaf. Otherwise, we keep dividing the region.

R-trees are mainly used in query and search for the geometric data analysis. The idea is to use a rectangle to cover all data points in a 2D space and then use smaller subrectangles to cover the points inside the parent rectangle. The subrectangles in the same level can overlap each other. See Fig. 2.5. The dimensions of each rectangle are not fixed. This allows the most flexibility in the partition of data in a space.

A popular problem in data science is as follows: Given a point location, find all of the rectangles or leaves that contain a chosen point [13, 42]. The idea of *R*-trees is to use the bounding rectangular boxes to decide whether or not we need to search inside the subtree. Therefore, most of the nodes in the tree are never read during a search. There are many applications that use *R*-trees in wireless networking. For instance, in queries for wireless broadcast systems, we need to quickly find a group of users in a heavy rain area to tell them that there may be a possibility of flooding. This is especially useful in social network systems. Social networking categorizes individuals into specific groups . Let us take a social network, such as *facebook*, that does not usually contain location parameters for users. When a local social event needs to find attendees in a hurry, they can ask *facebook* to send messages to local members. *R*-tree is the best way to identify the people to send this message.

2.6.4 NP-Hard Problems and Approximation of Solutions

Some problems are easy to solve such as sorting and searching problems. However, for other problems, even though we can find algorithms to solve the problem, there is no quick or fast algorithm to complete the task. Many of these require exponential time. If we want to have a precise answer, we may need to spend 100 years to solve this using a single machine. Cloud computing provides us the possibility of solving a difficult problem in a relatively short period of time. For instance, in RSA public cryptography, we should use the string (called key) with 2024 bits for encryption. This is because in distributed computing, a host computer links with thousands of other computers, which can break the encryption code if the key is shorter than a certain length. Both Bigdata and data science must consider security problems. However, this is not the main topic of this book.

Roughly speaking, an NP-hard problem would take years to run for a data set with a large input size. A problem is said to be an NP-problem if we can check the answer quickly, e.g. within $O(n)$ time. Many problems are found to have such a property. For instance, the traveling salesmen problem has this property. This problem involves finding a route for a traveler who plans to visit one city in each of the 50 United States exactly once. The question is, is there a route such that the total distance travelled is smaller than a given number K?

If we know the sequence of the list of 50 cites, then we can easily check to see if the answer is correct. If we do not know the answer, then we may need to spend 10 years in finding it [1, 12].

It was proven that the traveling salesmen problem is the hardest of the NP-problems, called an NP-complete problem. It means that if the traveling salesmen problem can be solved in polynomial time, then all NP-problems can be solved in polynomial time with regular computers. There is a famous unsolved problem in mathematics and computer science called the $P = ?NP$ problem. Here, P means that the problem can be solved in polynomial time by regular computers (for example, a Turing machine) and NP means that the problem can be solved in polynomial time by a non-deterministic Turing machine. NP-hard problems mean the problems that are harder than NP-complete problems [12, 21, 25].

Some data science problems are NP-hard. For these problems that are difficult to solve, we may need to find an approximation of the solution. In [40], there are many methods that can be used in finding a near best answer for NP-hard problems.

2.7 Online Searching and Matching

Searching and matching are two basic tasks over the Internet. Search is used to find an object in a set or space. For instance, if we are searching for a number in an integer set, we know the binary search method is the fastest for a sorted array of integers. Matching is usually try to fine a substring in a text that matches a given pattern that is also a string. Today, the image especially the face image matching is very popular. We will discuss the image problems in Chap. 5 [9, 11, 15, 16, 20].

2.7.1 Google Search

Google search is an excellent example for online search using mathematical algorithms [4, 9, 29]. We say that the algorithm is a mathematical algorithm since it uses some deep knowledge of mathematics not only the complexity from computer science point of view.

Google search (also called *PageRank*) seeks to establish a link graph and then calculates the importance of each web node (page). To explain this method, we start with the adjacency matrix of the page link graph, which is a directed graph. This matrix, shown in Fig. 2.6, is as follows [9]:

$$M = \begin{bmatrix} 0 & 1 & 1 & 1 & 0 \\ 1 & 0 & 1 & 0 & 0 \\ 1 & 0 & 0 & 0 & 0 \\ 0 & 0 & 0 & 0 & 1 \\ 0 & 1 & 1 & 0 & 0 \end{bmatrix}$$

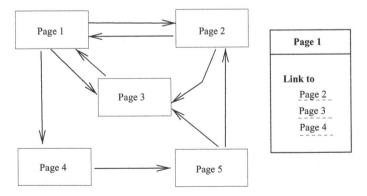

Fig. 2.6 Example of a Web link graph [9]

Its weight graph with the distribution of the average contribution to each of outgoing nodes is shown below. For instance, the first page "Page 1" gives $1/3$ of the contributions to each of its pointed neighbors: Page2, Page3, and Page4. We have the weight matrix:

$$W = \begin{bmatrix} 0 & \frac{1}{3} & \frac{1}{3} & \frac{1}{3} & 0 \\ \frac{1}{2} & 0 & \frac{1}{2} & 0 & 0 \\ 1 & 0 & 0 & 0 & 0 \\ 0 & 0 & 0 & 0 & 1 \\ 0 & \frac{1}{2} & \frac{1}{2} & 0 & 0 \end{bmatrix}$$

The transpose of W will be the matrix we are interested in, denoted by M_{PR}. Such a representation is very intuitive.

On the other hand, Page and Brin have used the following formula to rank the importance of each web page [4] called the PageRank of a page A:

$$PR(A) = \frac{(1-d)}{N} + d(PR(T_1)/C(T_1) + \ldots + PR(T_n)/C(T_n)), \tag{2.5}$$

where N is the total number of pages considered, $PR(T_i)$ is the PageRank of page T_i (which links to page A), $C(T_i)$ is the number of links going out of page T_i, and $d \in [0, 1]$ is a damping factor which is usually set to 0.85.

A simple algorithm runs the above formula in an iterative manner. It stops when an error limit is met.

It is astonishing when we check the relationship between the matrix M_{PR} and $PR(A)$. Let vector $x = (1/N, \cdots, 1/N)$. x^T be the transpose of x. We will then get another vector $M_{PR}x^T$ and so on for $M_{PR}^k x^T$. We know that $M_{PR}^k x^T$ will converge to a vector when k is big enough. Then

$$M_{PR}^{k+1} x^T = M_{PR}[(M_{PR})^k x^T] = (M_{PR})^k x^T.$$

Therefore, in such a case, $y = (M_{PR})^k x^T$ is the eigenvector of M_{PR}. After we add the dumping factor to the matrix, we have

$$G = \begin{bmatrix} \frac{(1-d)}{N} \\ \cdots \\ \frac{(1-d)}{N} \end{bmatrix} + d \cdot M_{PR}. \tag{2.6}$$

The eigenvector of G will be approximately

$$\begin{bmatrix} PR(A_1) \\ \cdots \\ PR(A_N) \end{bmatrix}.$$

2.7.2   Matching

Matching usually means searching for an object that may contain several elements in a set. The set may hold a structure, meaning that the elements in the set have relations and connections. Finding a substring in a DNA sequence is a good example. More profound research in finding a protein structure is a geometric problem.

The most useful matching technique today is still string matching, it is to find a certain substring from a big text file. A fast algorithm called the KMP algorithm is often used [12].

If the set is a random set, then the search will be trivial since we can only compare the elements from the set one by one. When the set contains a structure, such as an order, the search may use properties of the structure. In space, the set could be a topological structure such as components, a geometric structure such as distance metrics, or an algebraic relation such as rules.

Matching could also mean that we can find a partial or best match. For instance, finger print matching [20, 39].

2.7.3   High Dimensional Search

Recent research shows great interest in high dimensional search. This search is usually related to the nearest neighbor method. The fast algorithm is based on preexisting data structures. To find a new data point in a set of points, usually in the form of an n-dimensional vector, we want to find the point that is closest to the new point. The R-tree data structure was previously implemented in order to perform a fast search. This application is important in wireless networking. For instance, we have a number of wireless tower stations that cover cellular phone user communication. When a new phone user joins in the area, we need to locate the new phone and appoint a station to communicate with it. To search the cell phone without calculating its distance from all the stations, we will need a special data structure. The most popular today is the R-tree.

Another hot research topic due to the need of BigData applications is called subspace recovery. The following is an interesting question that has recently received much attention:

Thinking about a simple problem, there are 100 sample points on a 2D plane. We want to find a line that contains most of these sample points. Is this problem NP hard? The decision problem can be described as follows: Is there a line that contains at least 50 points? What about a curve instead of a line? This problem is highly related to manifold learning where we try to determine a cloud point set that represents a manifold such as a curve. The data points are not only 50–100, but they are also 50–100 GB. How do we use cloud computing technology to solve this problem?

The general version of this problem is called robust subspace recovery, which relates to dimension reduction [14, 18]: Given a collection of m points in R^n, if many but not necessarily all of these points are contained in a d-dimensional subspace T, can we find m? The points contained in T are called inliers and the remaining points are outliers. This problem has received considerable attention in computer science and statistics [14, 35, 38, 41]. However, efficient algorithms from computer science are not robust to adversarial outliers, and the estimators from robust statistics are difficult to compute in higher dimensions [2].

The problem is finding a T dimensional space that contains most of the points or a given ratio of inner points. How do we determine T in the fastest way? Does T have some sort of boundary?

Geometric search, especially high dimensional search, is a hot topic related to data science, along with incomplete data search, which is related to artificial intelligence.

2.8 Remarks: Relationship Among Data Science, Database, Networking, and Artificial Intelligence

In short, data science deals with massive data sets in different formats. The data sets are stored in databases or obtained over the Internet. Data science mainly uses statistical and artificial intelligence methods today. Fast algorithms and parallel and distributed algorithms are important in applications. Today, in terms of cloud computing technology, some software systems such as Hadoop and Spark are available to use [15, 22, 28]. These are essentially developed based on the principle of distributed computing. In the following chapters, we will study detailed methods for incomplete relations among data sets and machine learning. We will also discuss topological methods [5, 6, 10, 19] and advanced statistical methods, in Part II of this book, and other advanced topics and problems in Part III.

References

1. A.V. Aho, J.E. Hopcroft, J.D. Ullman, *The Design and Analysis of Computer Algorithms* (Addison-Wesley, Boston, 1974)
2. M. Belkin, P. Niyogi, Laplacian eigenmaps for dimensionality reduction and data representation. Neural Comput. **15**(6), 1373–1396 (2003)
3. B. Bollobas, *Random Graphs* (Academic, London, 1985)
4. S. Brin, L. Page, The anatomy of a large-scale hypertextual Web search engine. Comput. Netw. ISDN Syst. **30**, 107–117 (1998)
5. G. Carlsson, A. Zomorodian, Theory of multidimensional persistence. Discret. Comput. Geom. **42**(1), 71–93 (2009)
6. L. Cayton, Algorithms for manifold learning. Technical Report CS2008-0923, UCSD (2005)
7. L. Chen, *Discrete Surfaces and Manifolds: A Theory of Digital-Discrete Geometry and Topology* (SP Computing, Rockville, 2004)
8. L. Chen, *Digital Functions and Data Reconstruction* (Springer, New York, 2013)

9. L.M. Chen, Digital and Discrete Geometry: Theory and Algorithms, NY Springer (2014)
10. L. Chen, Y. Rong, Digital topological method for computing genus and the Betti numbers. Topol. Appl. **157**(12), 1931–1936 (2010)
11. L. Chen, H. Zhu, W. Cui, Very fast region-connected segmentation for spatial data: case study, in *IEEE Conference on System, Man, and Cybernetics* (2006). pp. 4001–4005
12. T.H. Cormen, C.E. Leiserson, R.L. Rivest, *Introduction to Algorithms* (MIT Press, Cambridge, 1993)
13. M. Demirbas, H. Ferhatosmanoglu, Peer-to-peer spatial queries in sensor networks, in *Third International Conference on Peer-to-Peer Computing*, Linkoping (2003)
14. D.L. Donoho, C. Grimes. Hessian Eigenmaps: new locally linear embedding techniques for high-dimensional data. Technical Report TR-2003-08, Department of Statistics, Stanford University (2003)
15. Afrati Foto and Jeffrey Ullman, (2009) Optimizing Joins in a Map-Reduce Environment. Technical Report. Stanford InfoLab. (2009)
16. K. Fukunaga, L.D. Hostetler, The estimation of the gradient of a density function, with applications in pattern recognition. IEEE Trans. Inf. Theory **21**(1), 32–40 (1975)
17. F. Harary, *Graph Theory* (Addison-Wesley, Reading, 1969)
18. M. Hardt, A. Moitra. Algorithms and hardness for robust subspace recovery, in *COLT*, pp. 354–375 (2013)
19. R. Ghrist, Barcodes: the persistent topology of data. Bull. Am. Math. Soc. **45**(1), 61–75 (2008)
20. R.C. Gonzalez, R. Wood, *Digital Image Processing* (Addison-Wesley, Reading, 1993)
21. J. Goodman, J. O'Rourke, *Handbook of Discrete and Computational Geometry* (CRC, Boca Raton, 1997)
22. J. Han, M. Kamber, *Data Mining: Concepts and Techniques* (Morgan Kaufmann, San Francisco, 2001)
23. H. Homann, *Implementation of a 3D Thinning Algorithm* (Oxford University, Wolf-son Medical Vision Lab., Oxford, 2007)
24. F.V. Jensen, *Bayesian Networks and Decision Graphs* (Springer, New York, 2001)
25. T. Kanungo, D.M. Mount, N. Netanyahu, C. Piatko, R. Silverman, A.Y. Wu, A local search approximation algorithm for k-means clustering. Comput. Geom. Theory Appl. **28**, 89–112 (2004)
26. R. Klette, A. Rosenfeld, *Digital Geometry, Geometric Methods for Digital Picture Analysis*. Computer Graphics and Geometric Modeling (Morgan Kaufmann, San Francisco, 2004)
27. T.C. Lee, R.L. Kashyap, C.N. Chu, Building skeleton models via 3-D medial surface/axis thinning algorithms. Comput. Vis. Graphics Image Process. **56**(6), 462–478 (1994)
28. T.M. Mitchell, *Machine Learning* (McGraw Hill, New York, 1997)
29. L. Page, S. Brin, R. Motwani, T. Winograd, The PageRank citation ranking: bringing order to the web. Technical Report. Stanford InfoLab. (1999)
30. T. Pavilidis, *Algorithms for Graphics and Image Processing* (Computer Science Press, Rockville, 1982)
31. W.H. Press, et al., *Numerical Recipes in C: The Art of Scientific Computing*, 2nd edn. (Cambridge University Press, Cambridge, 1993)
32. X. Ren, J. Malik, Learning a classification model for segmentation, in *Proceedings of the IEEE International Conference on Computer Vision*, pp. 10–17 (2003)
33. A. Rosenfeld, and A.C. Kak, *Digital Picture Processing*, 2nd edn. (Academic, New York, 1982)
34. S. Russell, P. Norvig, *Artificial Intelligence: A Modern Approach*, 3rd edn. (Pearson, Boston, 2009)
35. L.K. Saul, S.T. Roweis, Think globally, fit locally: unsupervised learning of low dimensional manifolds. J. Mach. Learn. Res. **4**, 119–155 (2003)
36. H. Samet, *The Design and Analysis of Spatial Data Structures* (Addison Wesley, Reading, 1990)
37. J. Shi, J. Malik, Normalized cuts and image segmentation. IEEE Trans. Pattern Anal. Mach. Intell. **22**(8), 888–905 (2000)

38. J.B. Tenenbaum, V. de Silva, J.C. Langford, A global geometric framework for nonlinear dimensionality reduction. Science **290**, 2319–2323 (2000)
39. S. Theodoridis, K. Koutroumbas, *Pattern Recognition* (Academic, Boston, 2003)
40. D.P. Williamson, D.B. Shmoys, *The Design of Approximation Algorithms* (Cambridge University Press, Cambridge, 2011)
41. Z. Zhang, H. Zha, Principal manifolds and nonlinear dimension reduction via local tangent space alignment. SIAM J. Sci. Comput. **26**(1), 313–338 (2005)
42. B. Zheng, W.-C. Lee, D.L. Lee. Spatial queries in wireless broadcast systems. Wirel. Netw. **10**(6), 723–736 (2004)

Chapter 3
Relationship and Connectivity of Incomplete Data Collection

Li M. Chen

Abstract A basic problem in data science is to classify a massive data set into different categories. In addition, when more new data samples are collected or come from online sources, we want to know how to put the new data sample into an appropriate category. This problem contains such information as classification, learning, and reconstruction. As we discussed in Chap. 2, the problem is central to several disciplinary areas including artificial intelligence, pattern recognition, and statistics. It is obvious that this is still a main problem in data science but the size of the data set has been changed to "BigData" and the ultimate goal of the process has become online based in many cases.

In this chapter, we introduce a comprehensive method for this complex problem. This method provides a unified framework for concurrent data science to solve the related problem. The method begins by considering data relations and connectivity among data points or data groups. These relations and connectivity are mostly incomplete. Then, we build a model to analyze, classify, or reconstruct the data or data sets, which is called the λ-connectedness method.

The method is built on general graphs. It also has some relations to persistent analysis that will be discussed in Chap. 6. Persistent analysis tries to find topological structures in cloud data. Combining λ-connectedness with persistent analysis can result in the classification of data sets into geometrically meaningful components, such as desired shapes or having characteristics of tracking and recognition.

3.1 Current Challenges of Problems in Data Science

To solve a problem related to a massive data set, we face the following challenges: (1) methodology challenges, (2) efficiency challenges, and (3) display visualization challenges.

Methodology Challenges If the data is in different forms from different resources, we cannot use a single method to solve it. A unified method may not be practical for

L.M. Chen (✉)
The University of the District of Columbia, Washington, DC, USA
e-mail: lchen@udc.edu

© Springer International Publishing Switzerland 2015
L.M. Chen et al., *Mathematical Problems in Data Science*,
DOI 10.1007/978-3-319-25127-1_3

solving a set of problems and may not even be practical for a single problem if input data is from different places. Multiple methods for multiple data types or problems are required, and a unified method for combined decision making may be selected (even from a collection of methods pool).

Efficiency Challenges Algorithms for cloud computing, networking based and MapReduce computing model. New algorithm design includes existing algorithm modification for large volumes of data in fast processing requests. For instance, k-mean is one of the best algorithms for unsupervised learning. However, k-mean is an exponential algorithm in that the error-clip level is very small. How do we use a modified k-mean with the approximation where part of the data is set to control a small error or be a small scaling factor and part of the data is set to be on a bigger scale.

DisplayVisualization Challenges How do we visualize the data and results for interactive interpretation? We can instantly input the human decision before the process is complete. However, what are the challenges? The human's power of imagination has been challenged. Interpretations include how we display the most important parts of the data or display the data from different "angles" for different users. In this book we mainly focus on methodology and efficiency challenges.

In general, the problems in current data science cannot be solved by classical methods (without modification). This is because many of the methods in statistics, pattern recognition, and artificial intelligence only work for local and relatively small sets of data. On the other hand, algebraic methods, including mathematical logic, can be used to find rules and formulas and topological geometric methods can be used to find properties of shapes for data sets. However, these methods meet difficulties in terms of effectiveness when the size of the data is very large.

How do we process a problem that is so large? For example, we can make a car inside of a factory house, but how do we build an aircraft carrier in a house? In such a case, what we want to do is build parts (big components) in different factories, and then we want to assemble them in an open area. This example provides us with a strategy. We could distribute the sub-problems to different machines or work stations (the Map procedure in MapReduce), and then after each subproblem has been solved, we can combine them into a whole. Now, when you combine the reduce procedure in MapReduce, you need to connect them together, not only putting the data together but also connecting them by melting electronic wires, plumbing, and polishing. It is not a simple task to combine the discrete objects into one. Each connection is measured by the connectedness of each feature (a parameter) similar to electrical wiring or plumbing. Therefore, the total quality of assembling or joining the components can be measured by the evaluation of weighted functions for all parameters.

Example 3.1. A simple and practical example is if we have a very large volume of data in 3D and would like to find the geometric structure of the data. Data that we would like to split into 8 computers (octree structure), where each process is 1/8 of the total, is similar to a large coconut. When data segmentation is down, each of the eight parts has three layers: the shell, coconut-flesh, and juice. When we put the results together, what do we have? For simple answer, we can say that there are three components, but for an accurate answer, the merge process requires much work by gluing the boundary together to predict the topological structure of the data. The ball of juice, the sphere of the shell, and a 3D manifold with a 3D hole.

There are still two problems: (1) How do we proceed when this data volume is gray scale or color images? (2) How do we design the algorithm when the gluing process must be implemented on different computers since a single computer cannot hold the entire volume? A clue is that we can pass the seed point from one computer to another. We will discuss the detailed algorithm later.

In this chapter, we introduce a systematic way for modeling this complex problem. To summarize, connectivity and connectedness should be used in such a case. This chapter will introduce a methodology for certain problems in similar cases.

3.1.1 Relations and Connectedness

Relation usually attempts to answer the question about how two things are related, such as similarity or difference. For instance, we say that two feature vectors that usually describe two objects are similar if the distance (or difference) between the two is small. Each individual sample or vector is treated the same. Classification is usually done in a similar way. On the other hand, connectivity, or connectedness, is a special relation that not only considers the difference or similarity of two individual, but also seeks to find a path or channel between the two objects.

Connectivity is a type of profound relation that tries to determine whether pair-wise relations are "linked." Connectedness is a measure for the degree of connectivity.

3.1.2 Multiple Scaling and Multiple Level Sampling

Multiple scaling and multiple level sampling are often used as a local importance is found. For example: When we drive from Washington DC to Los Angeles, there is no need to check the GPS every minute since we drive on a highway most of the time. However, we do need directions leaving and entering a city and when we stop for food. Multiple scaling and multiple level sampling are only needed in some locations while driving. Most of the time we only care about the highway road condition. We can put the GPS on sleep mode for the rest. This model is good for auto drive control. We do not need to store too many video images while on the highway. Therefore, we only need to analyze as necessary.

For any type of applications, there is a consideration of the rough treatment vs. the detailed treatment. The goal is to find the optimization of an outcome. Multiple scaling and multiple level sampling can be used as learning and evaluation techniques.

3.1.3   Relationship and Connectivity Among Data

One of the most important aspects of data science is measuring the relationship of two or more data sets. We can use the average values of data sets or shapes of data sets as measures. We can also use the size of data sets.

Mathematically, each of these characteristics can be a component in a feature vector of the data sets. The relationship should be described as a function $C(A, B)$ that is mainly the relationship measure of two feature vectors of set A and set B. The general model of such a relationship measure is a weighted similarity measure on vectors. The easiest method is to use the distance of these two vectors in the similarity measure. For instance, the following formula is a similarity measure [42].

$$C(A, B) = f(v_A, v_B) = 1 - \frac{d(v_A, v_B)}{\max\{\|v_A\|, \|v_B\|\}} \tag{3.1}$$

where v_S is the feature vector of set S. Therefore, the smaller difference is the bigger similarity.

The similarity is very general for relations. It is usually used in classifying individual vectors into different categories, as we discussed in Chap. 2, as k-NN and k-mean methods.

When we deal with a geometric problem, the location information is involved. The relationship between data adds a significant dimension, which is connectivity. To classify a set that considers if two data points are connected or near connected is very different from treating them as individual points. They are no longer independent events in terms of probability. For instance, in an image, two adjacent points can be considered as belonging to an object if they have similar intensity values. They may not be related if the two points are far away from each other even if they have the same value.

3.2   Generalized λ-Connectedness

In this section, we introduce a general model for the problems related to similarity measures for a domain that is a metric space. A metric space can be a graph or a Euclidean space. Here, we mainly present the method for graphs, which is called the λ-connectedness method [4, 6–8, 11, 14].

The λ-connectedness method is a simple system $< G, \rho >$, where $G = (V, E)$ is a graph and ρ is called a potential function. If $< G, \rho >$ is an image, then G is a 2D or 2D grid space and ρ is an intensity function of the image. For a color image, we can use $f = (red, green, blue)$ to represent ρ.

If $< G, \rho >$ represents the relationships among companies, then each vertex would be a company and each edge would be the existing business transactions between these two companies. In general, G can either be a directed or undirected graph, but we usually refer to G as an undirected graph. ρ can be a feature vector that contains information on size, revenue, profit, etc. of the company.

From a management prospective, $< G, \rho >$ could represent the individual relationships. Suppose G is a directed graph. In this case, each vertex represents a person, an edge (or arc in a directed graph) represents a supervisory relationship, and ρ represents the duties and powers in a company. ρ may also be a vector in which each component of the vector indicates a person's responsibilities in a specific aspect.

λ-connectedness can also be defined as a multiple level complex system. When decomposition is first completed, we can build a second λ-connectedness on the partitioned data domain [8, 14]. This is similar in some ways to the big-pixel technology in modern image segmentation (see Chap. 5) [32, 39]. This process can be iterative.

3.2.1 λ-Connectedness on Undirected Graphs

λ-connectedness can be defined on an undirected graph $G = (V, E)$ with an associated (potential) function $f : V \to R^m$, where R^m is the m-dimensional real space. Given a measure $\alpha_\rho(x, y)$ on each pair of adjacent points x, y based on the values $\rho(x), \rho(y)$, we define

$$\alpha_\rho(x, y) = \begin{cases} \mu(\rho(x), \rho(y)) & \text{if } x \text{ and } y \text{ are adjacent} \\ 0 & \text{otherwise} \end{cases} \tag{3.2}$$

where $\mu : R^m \times R^m \to [0, 1]$ with $\mu(u, v) = \mu(v, u)$ and $\mu(u, u) = 1$. Note that we can define $\mu(u, u) = c$ and $\mu(u, v) \leq c$ where $c \in [0, 1]$ for all u. α_ρ is used to measure "neighbor-connectivity." The next step is to develop path-connectivity so that λ-connectedness on $< G, \rho >$ can be defined in a general way.

In graph theory, a finite sequence $x_1, x_2, \ldots, x_n$ is called a path if $(x_i, x_{i+1}) \in E$. A path is called a simple path if $x_i \neq x_j$, $i \neq j$ excepting $x_1 = x_n$. The path $x_1, x_2, \ldots, x_{n-1}, x_n = x_1$ is called a cycle. The path-connectivity β of a path $\pi = \pi(x_1, x_n) = \{x_1, x_2, \ldots, x_n\}$ is defined as

$$\beta_\rho(\pi(x_1, x_n)) = \min\{\alpha_\rho(x_i, x_{i+1}) | i = 1, \ldots, n - 1\} \tag{3.3}$$

or

$$\beta_\rho(\pi(x_1, x_n)) = \prod\{\alpha_\rho(x_i, x_{i+1}) | i = 1, \ldots, n - 1\} \tag{3.4}$$

Finally, the degree of connectedness (connectivity) of two vertices x, y with respect to ρ is defined as:

$$C_\rho(x, y) = \max\{\beta(\pi(x, y)) | \pi \text{ is a (simple) path.}\} \tag{3.5}$$

For a given $\lambda \in [0,1]$, points $p = (x, \rho(x))$ and $q = (y, \rho(y))$ are said to be λ-connected if $C_\rho(x, y) \geq \lambda$. In image processing, $\rho(x)$ is the intensity of a point x and $p = (x, \rho(x))$ defines a pixel. (Note: We usually avoid saying that x and y are λ-connected because it does not contain the information of ρ. We may say that two pixels are λ-connected, but generally not that two points are λ-connected.)

If $< G, \rho >$ is an image, then this equivalence relation can be used for segmentation, i.e., partitioning the image into different objects. On the other hand, if a potential function f is partially defined on G, then one can fit f to ρ such that $< G, \rho >$ is λ-connected on G.

3.2.2 λ-Connectedness on Directed Graphs

The only difference between a directed graph $G = (V, A)$ and an undirected graph $G = (V, E)$ is that if $(x, y) \in E$, then $(y, x) \in E$, whereas if $(x, y) \in A$, then it may or may not be true that $(y, x) \in A$. Therefore, for λ-connectedness, there may be a case that p can be λ-connected to q but q cannot be λ-connected to p. For example, an irrigation system can be described on a directed graph. Suppose there are N water reservoirs in the system. Then, let x be a reservoir and $\rho(x)$ be the amount of energy/power of x. If $(x, \rho(x))$ is λ-connected to $(y, \rho(y))$, then reservoir x can transfer water to reservoir y. It may not be the case that $(y, \rho(y))$ can be λ-connected to $(x, \rho(x))$.

Similarly, define α_ρ and β_ρ on $< G = (V, A), f >$,

$$\alpha_\rho(x, y) = \begin{cases} \mu(\rho(x), \rho(y)) & \text{if arc } (x, y) \text{ is in A} \\ 0 & \text{otherwise.} \end{cases} \tag{3.6}$$

In a directed graph, a finite sequence $x_1, x_2, \ldots, x_n$ is called a strongly connected path if $(x_i, x_{i+1}) \in A$ for all $i = 1, \ldots, n - 1$. The sequence is called weakly connected if $(x_i, x_{i+1}) \in A$ or $(x_{i+1}, x_i) \in A$ for all i. The path-connectivity of a path $\pi = \pi(x_1, x_n) = \{x_1, x_2, \ldots, x_n\}$ is still defined as

$$\beta_\rho(\pi(x_1, x_n)) = \min\{\alpha_\rho(x_i, x_{i+1}) | i = 1, \ldots, n - 1\} \tag{3.7}$$

or

$$\beta_\rho(\pi(x_1, x_n)) = \prod\{\alpha_\rho(x_i, x_{i+1}) | i = 1, \ldots, n - 1\} \tag{3.8}$$

Thus, the degree of connectedness (connectivity) is defined as

$$C_\rho(x, y) = \max\{\beta(\pi(x, y)) | \pi \text{ is a (simple) path}\} \tag{3.9}$$

where the path can be a strongly connected path or a weakly connected path. A weakly connected path appears to be the same as a path in an undirected graph. In directed graphs, a path usually means a strongly connected path.

We should note that having λ-connectedness on undirected and directed graphs applies to different categories of problems. The former mainly deals with the similarities of two individuals on a "network." The latter is for "sending," "receiving," and "finding." In most applications, an acyclic directed graph is required for simplifying analysis procedures. An acyclic graph means that there is no (strongly connected) cycle in the graph [24].

3.2.3 Potential Function ρ and the Measure μ

For λ-connectedness, the potential function ρ and μ are two key factors. If G is a lattice or a ϵ-net [23], ρ may be a polynomial function. For a color image, ρ is the $RGB = (r, g, b)$ function. For a resource network, $\rho(x)$ can represent the utility function or resource potentials of the nodes in multi-dimensions. For example, a city that utilizes high electricity, generates more jobs, and has many universities can be represented by the following utility vector

$$\rho(theCity) = (highElectricity, moreJobs, manyUniversities, \ldots\ldots).$$

Measure μ may also vary depending on different situations. In previous work, three kinds of μ were used [4, 12]:

$$\mu_1(u, v) = \begin{cases} 1 - \frac{\|u-v\|}{\|u\|+\|v\|} & \text{if} \|u\| + \|v\| \neq 0 \\ 1 & \text{otherwise} \end{cases} \tag{3.10}$$

and

$$\mu_2(u, v) = 1 - \frac{\|u - v\|}{H}, \tag{3.11}$$

where $H = max\{\|u\| | u \in \rho(V)\}$. The third kind of μ is defined by a so-called gradual variation relation of a sequence of real (or rational) numbers $A_1 < A_2 < \ldots < A_n$. Here, $\rho : V \rightarrow \{A_1, A_2, \ldots, A_n\}$ for $G = (V, E)$ (or $G = (V, A)$). Let $u = A_i$ and $v = A_j$, $\mu(u, v) = 1$ if $i = j - 1, j$ or $j + 1$; otherwise, $\mu(u, v) = 0$. Thus, λ-connectedness based on gradual variation can be defined with the value of λ being "1." The gradually varied function ρ plays a key role in λ-connected fitting, which will be discussed in detail later. Rosenfeld used a similar concept called the "continuous" digital function for image processing [10, 33].

3.3 λ-Connected Decomposition and Image Segmentation

Partition a space based on a function on the space. A continuous or smooth looking part of the function can be viewed as a partitioned region. This is usually called a component or a connected region. In image processing, this is called a segmentation.

In mathematics, an example of contour maps is such a partition. The difference is that an image is not a continuous function in most cases. Contour maps only use clip-levels to separate functions, which is a hard cut (no tolerance).

Segmentation is a clustering method used in image processing [1, 18]. There are many kinds of segmentation, e.g., measurement space clustering, region growing, split-and-merge segmentation, edge detection, etc. Different types of data require different segmentation methods. For gray-scale images, such as velocity profiles, we can use region growing or edge detection. For texture images, such as waveform profiles or seismic sections, we may need rule-based segmentation.

As we know, a digital image F is a mapping from a grid-space Σ to the real set R (or R^n in general). Let S be a connected component of Σ. S is said to be uniform if F on S (a sub-image) has properties of uniformity. For instance, a popular uniformity measure is defined in terms of the maximum difference between any pixel value and the mean value of S.

Segmentation partitions an image into connected subsets (segments) [18, 20–22]. Each segment is uniform, and no union of adjacent segments is uniform. The formal definition of segmentation is then: In a digital image F, if there exists a non-empty segmentation $F_1, F_2, \ldots, F_m$, satisfying:

(1) $F_i \cap F_j = \emptyset$, if $i = 1, .., m, j = 1, ..m, i \neq j$,
(2) $\cup_{i=1,\ldots,m} F_i = F$,
(3) each F_i is connected,
(4) each F_i is "uniform," and
(5) if F_i and F_j are adjacent, then $F_i \cup F_j$ is not uniform,

then $\{F_1, F_2, \ldots, F_m\}$ is called a segmentation of F. In this section, we describe two geometrical segmentation methods, region growing and split-and-merge segmentation, and their relationship to λ-connectedness. For more information about image segmentation, see [26–28, 30].

3.3.1 λ-Connected Region Growing Segmentation

In a gray-scale image, intensity is the uniformity measure. A region (or segment) in an image may be viewed as a connected group of pixels, all with similar brightness. The region growing method begins with a single pixel, and then examines its neighbors to find a maximum sized connected region of similar pixels. In this manner, regions grow from single pixels. We can also use a region or grouped set of pixels as a seed instead of a single pixel. In this case, after selecting the partition (group of pixels), a uniformity test is applied to the region to see if it qualifies as a partition. If the test fails, the region is subdivided into smaller regions. This process is repeated until all regions are uniform. (The major advantage of using small regions rather than single pixels is that it reduces the sensitivity to noise.)

Region growing could form an equivalence relation to partition the image. λ-connected segmentation is used to partition the image by searching each λ-connected component in the image. The fast algorithm design technique such as depth-first-search or breadth-first-search can be used for implementation [9, 15]. This method has been applied in several different situations such as seismic layer search [4, 12], ionogram scaling [43], and tracking hurricane centers [25]. Ionogram scaling is a famous application of artificial intelligent in space data analysis [17, 44].

The algorithm is presented below using the breadth first search method. Breadth first search is a fast search approach to get a connected component on a graph. First, it starts at a vertex p and searches for all adjacent vertices. Then, it "inserts" all adjacent points (neighbors) into a queue. It "removes" a vertex from the queue and lets it be p before repeatedly finding p's neighbors until the queue is empty. Marking all of the vertices we visited, the marked vertices form a connected component.

We have also introduced this technology in Chap. 2 in the artificial intelligence search subsection. To modify this idea and use it in λ-connected search, we only need to check if two adjacent points are λ-adjacent. The modified algorithm is shown below:

Algorithm 3.1. Breadth first search technique for λ-connected components.
> Step 1: Let p_0 be a node in $< G, \rho >$. Set
> $$L(p_0) \leftarrow * \text{ and } QUEUE \leftarrow QUEUE \cup p_0$$
> i.e., labeling p_0 and p_0 is sent a queue $QUEUE$.
> Step 2: If $QUEUE$ is empty, then go to Step 4; otherwise,
> $$p_0 \leftarrow QUEUE \text{ (top of } QUEUE\text{). Then,}$$
> $$L(p_0) \leftarrow 0.$$
> Step 3: For each p with an edge linking to p_0 , if
> $$L(p) \neq 0, L(p) \neq *, \text{ and } C(p, p_0) \geq \lambda, \text{ then}$$
> $$QUEUE \leftarrow QUEUE \cup p \text{ and } L(p) \leftarrow *. \text{ Then, go to Step 2.}$$
> Step 4: Stop. $S = \{p : L(p) = 0\}$ is one λ-connected part.

We can easily tell that this algorithm is an optimum time algorithm.

Proposition 3.1. *Algorithm 3.1 uses breadth first search for λ-connected components and is a linear time algorithm.*

3.3.2 λ-Connected Split-and-Merge Segmentation

For 2D images, split-and-merge segmentation uses a quadtree partition of an image and, hence, is sometimes called quadtree segmentation. In this section, we only present the principle of the segmentation method. A detailed implementation will be discussed in Chap. 5.

This algorithm begins with tree nodes (representing square regions of the image) at some intermediate level of the quad-tree. If a square is found to be non-uniform, then it is replaced by its four son-squares (split). Conversely, if four son-squares are segmented and a region in a son-square can merge with a region in another son-square in terms of adjacency and uniformity, they will be merged until no more pairs of regions can be merged. These four son-squares are replaced by a single square (merge), called a segmented square. This process continues recursively until no further splits or merges are needed.

This advanced segmentation was first designed by Horowitz and Pavlidis in [19]. In their traditional split-and-merge segmentation, the uniformity is measured by the mean of the merged region which is a statistical measure. Horowitz and Pavlidis thought the algorithm should be a fast algorithm. The optimum algorithm for the split-and-merge method was due to a special data structure made by Chen in [4]. The algorithm arrived at $O(n \log n)$.

In λ-connected split-and-merge segmentation, we merge two regions into one if the merged region is λ-connected for any possible path in the region. Such a region is called a normal λ-connected set [4, 12]. This is because if we use one-path λ-connectedness, there is no need to use a quadtree since it is an equivalence relation. Split-and-merge segmentation only preserves reflexivity and symmetry and is not a mathematical partition or equivalence classification.

Let us present this algorithm with λ-connectedness as measures in the following [4, 12].

Algorithm 3.2. The split-and-merge technique for λ-normal-connectedness.

Step 1: Let Σ_2 be a $(2^l \times 2^l)$ array. $L \leftarrow \Sigma_2$. Let L be a set.

Step 2: If region L is already homogeneous, i.e. L is λ-normal-connected set, $NOMERGE \leftarrow NOMERGE \cup L$ and go to Step 3; otherwise, divide L into four subsquares L_{11}, L_{12}, L_{21}, and L_{22}; then $NOSPLIT \leftarrow NOSPLIT \cup \{L_{11}, L_{12}, L_{21}, L_{22}\}$.

Step 3: If there exists four regions which are four son-squares of some region Ls
in $NOMERGE$, namely L_{11}, L_{12}, L_{21}, and L_{22},
they will be combined by the merging approach. It tests any two λ-normal-connected subset, which are neighbors
and belong to different regions in L_{11}, L_{12}, L_{21}, and L_{22}, merge
these subsets if their union is a λ-normal-connected subset of L.
$NOMERGE \leftarrow NOMERGE - \{L_{11}, L_{12}, L_{21}, L_{22}\}$. Otherwise,
if there are no such four regions, then stop when $NOSPLIT$ is empty,
choose a region from $NOSPLIT$, send it to L, and go to Step 2.

Proposition 3.2. *Algorithm 3.2 using quadtrees for normal λ-connected components is a log-linear $O(n \log n)$ time algorithm.*

The region growing method was referred to as the component labeling method in Rosenfeld and Kak [18, 29, 34]. The technique was only used for binary images. If we set $\lambda = 1$, then λ-connected segmentation will obtain the same results as the component labeling method.

For split-and-merge segmentation, the λ-connectedness method is good for certain situations. As we described in Sect. 3.3, the λ-connected split-and-merge segmentation finds the part in which a pixel must be λ-connected with all of its neighbors. In other words, this part must be "continuous." Rosenfeld discussed the topic of "continuous" in digital pictures [33] where he treated the continuity as the difference not to be greater than 1. We can see that λ-connectedness in this case is more general. For more discussion, see [6, 10].

3.4 λ-Connectedness for Data Reconstruction

Data reconstruction means to fill missing data points based on some data samples. Data fitting is one of the mathematical treatments. In computer science, data recovery is probably more general comparing to data reconstruction, but it mainly refers to damaged or corrupted data. Data recovery can also mean to recovery a subset of data from a database. Netflix problem discussed in Chap. 1 was relatively a new problem in data recovery.

Interpolation and approximation are two basic data fitting methods. If all of the original points remain the same value after fitting, it is called interpolation; otherwise, it is called approximation. In mathematics, there are many fitting algorithms such as B-spline, Bezier polynomial, etc. [31]. When the entire data set for a region was obtained, we can modify some points, which have unusual values compared to their neighbors. This is called smoothing. Using convolution filters to eliminate high frequency points is a typical smoothing method [18].

Segmentation and fitting are two procedures that are opposites to each other. Segmentation finds the (connected) component in which each element has the same "property," so each component can be represented by an element in the component. Fittings find the (whole) distribution function based on given "representative" (sample) points. In the fitting process, if there is no sample point which was picked for a component, then the component will not appear after the fitting.

The NetFlix matrix completion problem (see Chap. 2) is a fitting problem. When a continuity of elements in the matrix is assumed, this is very reasonable assumption when we deal with a similar group of people with or without a partition that is needed. Such a problem will turn into data reconstruction, that is similar to data fitting especially for some undetermined models. There is no need to have $O(Nr \log N)$ limitations for the method developed. According to discrete harmonic functions and gradually varied fittings, we can always find a solution to fill the matrix when several entries are given. See [10]. Therefore, the NetFlix matrix completion problem could be solved in other ways. On the other hand, holes (zero elements) in the matrix might be real not a noise or missing data. In this case, we may need to use topological method to compute this problem. We will discuss it in Chap. 6.

3.4.1   λ-Connected Fitting

Generally, a λ-connected fitting can be described as follows: Given an undirected graph $G = (V, E)$ or a directed graph $G = (V, A)$ and a subset J of V, if $\rho_J : J \to R^m$ is known, then λ-connected fitting is to find an extension/approximation ρ of ρ_J such that $\rho : V \to R^m$ is λ-connected on G (meaning that $< G, \rho >$ has only one λ-connected component) for a certain λ [11, 13].

With the concept of λ-connected fitting defined, the questions to ask are whether or not the λ-connected fitting exists and how to determine the λ-connected fitting function?

We will discuss more about the existence of the function in the next subsection. Using the same strategy as the proof of the theorem on gradually varied fill given in [3], we arrive at:

Theorem 3.1. *Let G be a simple undirected graph or an acyclic directed graph. Assume that μ is given by (3.9) or (3.10). The necessary and sufficient condition under which there exists a λ-connected interpolation is that for any two vertices x and y in J, every shortest path between x and y in G is λ-connectable.*

In the theorem, λ-connectable for a path π means that there is a valuation to each vertex $a, a \in G - J$, on the path such that $\beta_\rho(\pi)$ is not less than λ. For the proof of this theorem, refer to [10].

The digitized method called gradually varied fitting is a special type of λ-connected fitting. Let D be a connected subset in V and $J \subset D$. If given $f_J : J \to \{A_1, A_2, \ldots, A_m\}$, is there an extension of $f_J, f_D : D \to \{A_1, A_2, \ldots, A_m\}$ such that for all $p \in J, f_J(p) = f_D(p)$? This is called the gradually varied interpolation problem. The existence theorem is stated below [3, 10]:

Theorem 3.2. *The necessary and sufficient condition under which there exists a gradually varied interpolation is that for any two points p and q in J, the length of the shortest path between p and q in D is not less than the level-difference between p and q.*

3.4.2   Intelligent Data Fitting

In some situations, if there are outlier points in the sampled data, traditional fitting methods may generate faulty results. In such a case, an advanced technique, called robust fitting (or regression), can be used if we assume that the data comes from the same object. If the data comes from different classes/objects, for example one data point is sampled in a sandstone layer and the other is sampled from a coal layer, fitting will damage the original information. What is more, in geophysical data processing, if we want to reconstruct a set of data between two points with the same depth and suppose there is a fault between these two points, the fitting will finally erase the fault. For this reason, a more intelligent data fitting technique is needed [13, 35, 37, 38].

Chen et al. described an intelligent data fitting method to reconstruct a velocity volume in [13]. The key idea is to perform a λ-connected segmentation first, then fit the data.

It is not very hard to implement λ-connected segmentation and to find its applications. λ-connected fitting, such as gradually varied fitting, has its own theorems, but it is more theoretical in nature.

There are some mathematical issues regarding λ-connected fitting: (1) λ-connected approximation if there is no λ-connected interpolation with respect to guiding points, (2) optimum fitting, and (3) λ value determination.

First, the λ-connected interpolation or gradually varied interpolation may not always exist. An approximation method can be used to get a λ-connected fitting in terms of optimal uniform approximation [5]. For least squared fittings, since it relates to derivatives, more tools and knowledge to deal with this problem are required.

Second, the λ-connected fitting may not be unique and as we can expect, the question then becomes, which fitting is best?

Basically, different answers derive from different criteria. More research on this issue will be pursued in the near future.

An alternative way to solve this problem is to perform a λ-connected fitting first because it can handle the irregular domain easily. Then, we can use a standard fitting method for the secondary process. For instance, we can first generate a gradually varied surface, then use the B-spline method to fit the surface when a smooth surface is required.

Third, calculating the value of λ in either segmentation or fitting is a critical issue. For segmentation, the selection of λ depends on how many segments we want to separate. In Sect. 3.5, we will discuss the maximum-connectivity spanning tree method, which can be used to find the value of λ. Since the process of finding the maximum-connectivity spanning tree is time consuming, one may use a histogram or an experimental test to solve the problem.

A binary "finding" method can be used to test the value of λ. First, test for $\lambda = 0.5$. If λ is too big, then $\lambda = 0.25$ will be tried. If this λ is too small, then $\lambda = 0.75$ will be tried, and so on. Similarly, for λ-connected fitting, the same method may also be used.

3.4.3 λ-Connected Fitting with Smoothness

A typical λ-connected fitting method can only reconstruct "continuous" surfaces or functions. If we want to fit a smooth surface or a surface with a certain order of smoothness, we need to consider the derivatives in the fitting algorithm.

In this section, we only present the simplest algorithm that will make a first order smooth function. For other cases, see [10, 11].

As we know, the advantage of λ-connected fitting is fitting a surface without a mathematical formula. It is usually needed for a data fitting such as B-spline. In addition, the domain is not required to be regular, for example, a rectangle.

It is important that we assume that there is a way to calculate gradients using $< D, f >$. In addition, we need an iteration process so we can determine what is the most fitted "fitting."

If D is two dimensional (it does not necessarily have to be a 2D domain), we know $f_J : J \to R$ and $\frac{df}{dv} : D \to R$. Then, the idea of an algorithm can be developed as follows. For a specific problem, we know how to calculate gradients based on the discrete values on D. For example, for a rectangular area, we can use the difference method to get derivatives.

Because at the beginning of the fitting only some of the f values are available, the gradient calculation is not finalized. An iteration process is designed for the approximation. It is easy to see that if the domain has its triangulation, we can perform a similar calculation.

Algorithm 3.3. λ-connected fitting with complete first-order gradients. Given $f_J : J \to R, J \subset D$ and $\frac{df}{dv} : D \to R$, and we want to get $f = f_D$.

Step 1   Check if f_J satisfies the λ-connected interpolation condition. We can assume that $\frac{df}{dv}$ is λ-connected. Start at a point p in J, get a point q from its neighborhood N_p. ($N_p = N(p, B)$, where B contains all adjacent vertices of p.)

Step 2   Obtain all $f(q)$ such that $f_{J \cup \{q\}}$ satisfies the λ-connected interpolation condition. Put all of $f(q)$'s into a set $F(q)$.

Step 3   Select an f_0 in $F(q)$ such that the gradient on q (which is only based on the existing set $f_{J \cup \{q\}}$) is the closest to $\frac{df}{dv}$ at point q. According to Theorem 3.2, $F(q)$ can be determined. For instance, we can use the difference analysis method to get the gradient from the values of the points [10, 11]. Searching for the best combination from all possible combinations for q, select a new neighbor of p. Repeat the process from Step 2 until all neighbors are fitted.

Step 4   Select the new guiding point p sequentially or randomly. Select a neighbor of p and go to Step 2, repeat until all points are fitted in D.

Step 5   For all points that have new values, calculate the gradients. It should be different from the first time the gradients were calculated since all values of f are now known. Find the best $f(p) \in F(p)$ such that the "actual" $\frac{df}{dv}$ (based on f) is the closest to the "ideal" gradient $\frac{df}{dv}$ known before the interpolation. f must remain λ-connected during the entire process.

3.5   Maximum Connectivity Spanning Tree and λ-Value

The value of λ determines the number of partitions in vertex set $< G, \rho >$. In this section, we like to use the idea of minimum spanning tree (MST) to find the best λ-value. In other words, the maximum connectivity spanning tree is introduced to provide the total information for what value of λ should be selected. This idea was first presented in [7].

We have introduced the spanning tree of a graph G in Chap. 2. The maximum connectivity spanning tree is the one in which there is a path in the tree that has the maximum connectivity for every pair of points.

In this section, we will still use the Kruskal's algorithm [15]. In order to find the maximum connectivity spanning tree, we first calculate all the neighbor-connectivities first for each adjacent pair in $< G, \rho >$ to form a weighted graph [8].

Algorithm 3.4. The maximum connectivity spanning tree algorithm. We modified Kruskal's algorithm here to find a maximum connectivity spanning tree, where $G = (V, E)$ is the original graph. The weight for each pair of adjacent points indicates the neighbor-connectedness [8].

Step 1: Let $T = V$.

Step 2: Repeat Steps 3–4 until T has $|V| - 1$ edges.

Step 3: Find an edge e with the maximum connectivity value.

Step 4: If $T \cup e$ does not have a cycle, then $T \leftarrow T \cup e$ and
delete e from G; otherwise, delete e from G. Go to Step 3.

One advantage of using the maximum connectivity spanning tree is that it gives the complete information for λ-connectivity. With this tree, we can easily get a refinement of λ-connected classification. We can also find which value of λ should be selected for a particular segmentation. For example, the number of segments that is desired is controllable. However, a disadvantage is that algorithm C takes $O(n^3)$ time in terms of computational complexity and requires extra space to store the tree. It is very slow when a large set of points/edges is considered.

After the maximum connectivity spanning tree is generated, we can easily find all connected components for each λ. Consider the example shown below. Suppose that one wants to find the general relationship among major cities across the USA based on size, population, and political and economic importance. We assume the graph and potential values of cities are shown in Fig. 3.1.

For simplicity, formula (3.11) is used to compute the neighbor-connectivity, i.e., $\mu = 1 - |u - v|/10$. Then, the connectedness on each edge can be calculated. The result is shown in Fig. 3.2.

According to Algorithm 3.4, the maximum connectivity spanning tree can be computed, shown in Fig. 3.3. When λ is set to be 0.8 or smaller, it is always to generate the same λ-connected component(s) that contains all cities. If λ is 0.9, then there are two λ-connected components: one contains San Francisco and Los

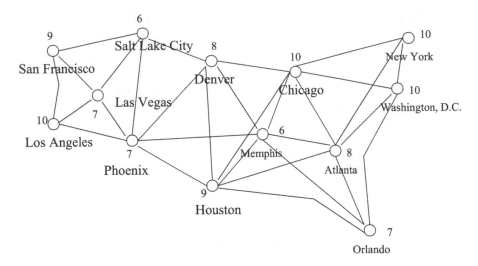

Fig. 3.1 Example of the potential values for major cities in the USA

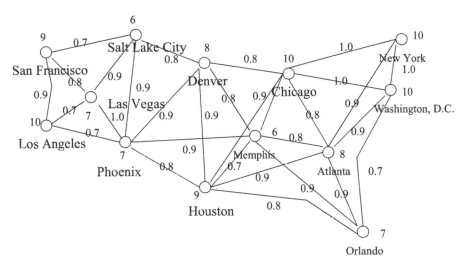

Fig. 3.2 The connectivity map based on the potential function

Angeles, and another one contains the rest of the cities. Therefore, a maximum connectivity spanning tree provides full information of the λ-connectedness among vertices.

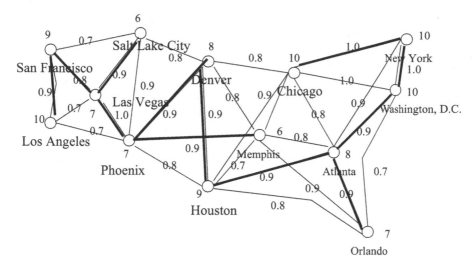

Fig. 3.3 The maximum connectivity spanning tree in Fig. 3.2

3.6 λ-Connectedness and Topological Data Analysis

Finding the topological structure of a massive data set has gained much attention in mathematics and engineering. Space data [36, 40] and meteorological data analysis require topological solutions.

A set of individual data points (cloud data) does not have a topological structure mathematically. However, since they are sampled from the real world, we can only get a discrete set of samples. So the human interpretation of each unique cloud data set means something different. How do we find the best interpretation that matches the original object or event [2, 46]?

We know each point is an independent component in sampling. However, we can interpret each data point as an area or volume (usually a disk or a ball), but we would not know how big the area or volume is. When the area is very small, the object may still remain not connected, but when the coverage area of a point is very big, the total area will be connected and become one connected object. So what would the radius be such that the filled area is the closest to the real world? This is also related to "data reconstruction" since a different radius chosen is the simple learning process.

While changing the value of radius, the homology or the number of holes does not always change, so the fit with the most unchanged holes is or might be referred to as the best fit. This method is called persistent homology analysis [2, 17, 47]. This method is becoming one of the major developments in topological data analysis.

In topology, homology groups usually indicate the number of holes in each dimension. Here, we treat this concept as a number of holes.

A λ-connected component indicates the topology. The homology of the λ-connected segmentation is the bar-code of persistent analysis. We only consider the original data points and do not need to work on the actual filling using disks. If a component contains a hole, we can use Minkowski's sum.

In the persistent homology method, we make a sample point grow in its volume with a radius r. When the radius r changes from 0 to a big number, the data will change from individual data points to a large volume until fills the entire space.

The persistent homology method calculates the homology groups (number of holes in each dimension) for each r. It would make some sense that the same topology (homology groups) that covers most of r will be the primary topological structure of M, the data set.

In Chap. 6, we will discuss more about this aspect.

3.7 Remark: Future Concerns

λ-connectedness has advantages in that it can work with both a search method with topological characteristics and a classification method related to geometric data classification [6, 11, 41].

It is a general methodology for multitasking, independent to variability in BigData, and a fast implementation method for computer programming due to Depth First Search and Breadth First Search technology.

To get the shape of a component in the data sets, we must consider other features, such as average value and shape type. In summary, connectivity can be applied to find the topological structure and λ-connectivity can be used to find classification. This classification in geometry is the shape of the object.

Learning in λ-connectedness includes learning the local lambda value, quadtree analysis, and the kernel method or dynamically moving kernel methods.

Homology can find topological structure, but we still need to find the geometric structure of data points. For this purpose, λ-connectedness is a good candidate for data science.

Topological analysis is needed to model data sets that may not fill the entire space. For instance, the rainfall during a hurricane will contain a centric area that does not have a lot of precipitation. Data reconstruction considers the topology of the data sets.

When we consider a specific problem, what features are the most important? Average value, shape, center location, or other. λ-connectedness will provide a general tool for these problems.

For any data set, with or without initial value, when a basic partition or classification is made, we can build a function related to each class. Further classification or decomposition of the original data will use this information. For instance, for the computation of homology groups in terms of persistent analysis or Morse theory, we can use the connectivity of the data on the previous partition classification . Not only do we need the simple distance metric but we also require

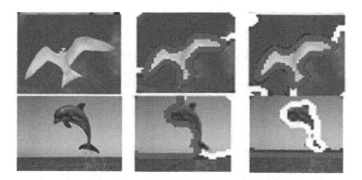

Fig. 3.4 Normalized cut combined with λ-connectedness: (**a**) original image, (**b**) normalized cut, (**c**) normalized cut combined with λ-connectedness (Color figure online)

additional information, though the distance metric (used in most persistent methods) is important. For instance, in circle packing, for the deformation of the data, we want to find the deformation trice. This trice will represent the customer's marketing tendencies.

The λ-connectedness method will provide more refined and detailed techniques to the modern problem related to BigData. Again, for existing techniques, the λ-connectedness method will be highly applicable to reexamine a practical problem related to data science in finding and modeling a detailed structure of the problem. It could bring us one step closer to the satisfactory or desired solution of the problem and can be used to articulate the problem [4, 32, 39, 44, 45].

For instance, Elhefnawy et al have used λ-connectedness with normalized cut in image segmentation, the result is very promising [16]. See Fig. 3.4.

References

1. B. Basavaprasad, R.S. Hegadi, A survey on traditional and graph theoretical techniques for image segmentation. Int. J. Comput. Appl. (Recent Adv. Inform. Technol.) **1**, 38–46 (2014)
2. G. Carlsson, Topology and data. Bull. (New Ser.) Am. Math. Soc. **46**(2), 255–308 (2009). doi:10.1090/s0273-0979-09-01249-x
3. L. Chen, The necessary and sufficient condition and the efficient algorithms for gradually varied fill. Chin. Sci. Bull. **35**, 10 (1990). *Abstracts of SIAM Conference on Geometric Design*, Temple, AZ, 1989
4. L. Chen, The lambda-connected segmentation and the optimal algorithm for split-and-merge segmentation. Chin. J. Comput. **14**, 321–331 (1991)
5. L. Chen, Gradually varied surface and its optimal uniform approximation (*IS&T/SPIE Symposium on Electronic Imaging*). SPIE Proc. **2182**, 300–307 (1994)
6. L. Chen, λ-connectedness and its application to image segmentation, recognition, and reconstruction. Ph.D. thesis, University of Luton, 2001
7. L. Chen, λ-connected approximations for rough sets, in *Lecture Notes in Computer Science*, vol. 2457 (Springer, Berlin, 2002), pp. 572–577

8. L. Chen, *Discrete Surfaces and Manifolds: A Theory of Digital-Discrete Geometry and Topology* (SP Computing, Rockville, 2004)
9. L. Chen, λ-measure for bone density connectivity, in *Proceedings of IEEE International Symposium on Industrial Electronics* Montreal, QC, pp. 489–494, 2006
10. L. Chen, *Digital Functions and Data Reconstruction* (Springer, Berlin, 2013)
11. L. Chen, O. Adjei, λ-connected segmentation and fitting: three new algorithms, in *Proceedings of IEEE conference on System, Man, and Cybernetics*, 2004
12. L. Chen, H.D. Cheng, J. Zhang, Fuzzy subfiber and its application to seismic lithology classification. Inform. Sci. Appl. **1**(2), 77–95 (1994)
13. L. Chen, D.H. Cooley, L. Zhang, An Intelligent data fitting technique for 3D velocity reconstruction. Appl. Sci. Comput. Intell. Proc. SPIE *3390*, 103–112 (1998)
14. L. Chen, O. Adjei, D.H. Cooley, λ-connectedness : method and application, in *Proceedings of IEEE conference on System, Man, and Cybernetics 2000*, pp. 1157–1562, 2000
15. T.H. Cormen, C.E. Leiserson, R.L. Rivest, *Introduction to Algorithms* (MIT Press, Cambridge, 1993)
16. W. Elhefnawy, L. Chen, Y. Li, Improving normalize-cut image segmentation using λ-connectedness, in Presented in the meeting of The Extreme Science and Engineering Discovery Environment (XSEDE), St. Louis, 2015
17. P. Ghosh, F.T. Berkey, Autonomous identification and classification of ionospheric sporadic E in digital ionograms. Earth Space Sci **2**(7), 244–261 (2015)
18. R.C. Gonzalez, R. Wood, *Digital Image Processing* (Addison-Wesley, Reading, MA, 1993)
19. S.L. Horowitz, T. Pavlidis, Picture segmentation by a tree traversal algorithm. J. ACM **23**, 368–388 (1976)
20. F. Jiang, M.R. Frater, M. Pickering, Threshold-based image segmentation through an improved particle swarm optimisation, in *Proceedings of International Conference on Digital Image Computing Techniques and Applications (DICTA)*, pp. 1–5, 2012
21. R. Klette, *A Concise Computer Vision* (Springer, Berlin, 2012)
22. D. Lazarevi, M. Misic, B. Cirkovi, Image segmentation as a classification task in computer applications, in *8th International Quality Conference*, 2014
23. S. Lefschetz, *Introduction to Topology* (Princeton University Press, Princeton, 1949)
24. C.L. Liu, *Elements of Discrete Mathematics* (McGraw Hill, New York, 1985)
25. C.-T. Lu, Y. Kou, J. Zhao, L. Chen, Detecting and tracking region outliers in meteorological data. Inform. Sci. **177**, 1609–1632 (2007)
26. A. Mouton, On artefact reduction, segmentation and classification of 3D computed tomography imagery in baggage security screening. Ph.D. thesis, University of Cranfield, 2014
27. D. Mumford, J. Shah, Optimal approximation by piecewise smooth functions and associated variational problems. Commun. Pure Appl. Math. **42**, 577–685 (1989)
28. N. Pal, S. Pal, A review on image segmentation techniques. Pattern Recogn. **26**, 1277–1294 (1993)
29. T. Pavilidis, *Algorithms for Graphics and Image Processing* (Computer Science Press, Rockville, MD, 1982)
30. J.-C. Pinoli, *Mathematical Foundations of Image Processing and Analysis*, vols. 1 and 2 (Wiley, New York, 2014)
31. W.H. Press et al., *Numerical Recipes in C: The Art of Scientific Computing*, 2nd edn. (Cambridge University Press, Cambridge, 1993)
32. X. Ren, J. Malik, Learning a classification model for segmentation, in *Proceedings of IEEE International Conference on Computer Vision*, pp. 10–17, 2003
33. A. Rosenfeld, "Continuous" functions on digital pictures. Pattern Recogn. Lett. **4**, 177–184 (1986)
34. A. Rosenfeld, A.C. Kak, *Digital Picture Processing*, 2nd edn. (Academic, New York, 1982)
35. S. Russell, P. Norvig, *Artificial Intelligence: A Modern Approach*, 3rd edn. (Pearson, Boston, 2009)
36. H. Samet, *The Design and Analysis of Spatial Data Structures* (Addison Wesley, Reading, MA, 1990)

37. L.K. Saul, S.T. Roweis, Think globally, fit locally: unsupervised learning of low dimensional manifolds. J. Mach. Learn. Res. **4**, 119–155 (2003)
38. A. Sengur, I. Turkoglu, M.C. Ince, A comparative study on entropic thresholding methods. Istanbul Univ. J. Electr. Electron. Eng. **6**(2), 183–188 (2006)
39. J. Shi, J. Malik, Normalized cuts and image segmentation. IEEE Trans. Pattern Anal. Mach. Intell. **22**(8), 888–905 (2000)
40. M. Spann, R. Wilson, A quad-tree approach to image segmentation which combines statistical and spatial information. Pattern Recogn. **18**, 257–269 (1985)
41. A. Teran, Real-time multi-target tracking: a study on color-texture covariance matrices and descriptor/operator switching. Ph.D. thesis, University of Paris Sud Paris XI, 2013
42. S. Theodoridis, K. Koutroumbas, *Pattern Recognition* (Academic, Orlando, FL, 2003)
43. L.C. Tsai, F.T. Berkey, Ionogram analysis using fuzzy segmentation and connectedness techniques. Radio Sci. **35**, 1173–1186 (2000)
44. M.Y. Wang, J.S. Ping, Martian ionogram scaling by the object tracking method. Sci. China Phys. Mech. Astron. **55**(3), 540–545 (2012)
45. M. Yamanaka, M. Matsugu, Information processing apparatus, control method for information processing apparatus and storage medium. US 8792725 B2, Canon Kabushiki Kaisha. http://www.google.com.ar/patents/US8792725
46. C.T. Zahn, Graph-theoretical methods for detecting and describing gestalt clusters. IEEE Trans. Comput. **20**(1), 68–86 (1971)
47. A.J. Zomorodian, *Topology for Computing*. Cambridge Monographs on Applied and Computational Mathematics (2005)

Part II
Data Science Problems and Machine Learning

Chapter 4
Machine Learning for Data Science: Mathematical or Computational

Li M. Chen

Abstract Machine learning usually requires getting a training and testing set of samples. The training set is used to obtain the model, and then, the testing set is used to verify the model. In general, a machine learning method requires an iterated process for reaching a goal. Machine learning is one of the research areas in artificial intelligence. Machine learning is mainly used to solve problems in classification and clustering. The distinction is that machine learning uses automated algorithms to learn from sample data for finding rules or making classifications. Therefore, the earliest machine learning method should be the regression method in statistics. Modern machine learning does not rely on a mathematical model such as linear equations used in regression. Online data mining and networking based applications nowadays request that machine learning be able to make decisions based on partial data sets. When more data samples are available, the algorithm must be able to adjust accordingly. Therefore, in cloud computing, and BigData related methods in data science, machine learning becomes the primary technology. We have introduced the PCA, k-NN and k-means, and other methods in artificial intelligence in Chap. 2. In this chapter, we will do an overview of other important machine learning methods such as decision trees, neural networks, and genetic algorithms. We will also introduce variational learning, support vector machine, and computational learning theory with some problems related to mathematical data processing.

4.1 Decision Trees and Boosting Process

Decision tree technology is one of the most flexible methods in machine learning adapted from pattern recognition [27, 29]. Theoretically, the decision tree method can solve every problem. Let us look at the following example: A university would admit a high school student if he or she has (a) An SAT score above 2300, (b) Community service hours (CSH) above 500, (c) A good essay (ESS), (d) An

L.M. Chen (✉)
The University of the District of Columbia, Washington, DC, USA
e-mail: lchen@udc.edu

© Springer International Publishing Switzerland 2015
L.M. Chen et al., *Mathematical Problems in Data Science*,
DOI 10.1007/978-3-319-25127-1_4

excellent recommendation (REC), (e) A GPA above 3.85, and (e) Some special talents in sports, music, or art (TAL).

We form a feature vector similar to $(SAT, CSH, ESS, REC, GPA, TAL)$. A decision can be made based on whether the students satisfy the above criteria. However, if student John has SAT 2400 but 350 hours of community service, would the school still consider him? If so, the decision-making procedure would look like a tree, a decision tree. The decision tree is usually a binary tree. We can assume that the left node indicates failure and the right node indicates success. Therefore, we have two categories.

The first node at Level 1 will check for the SAT score, which John passed. Then, going to the right son-node, we would check for CSH, which John failed. We would then recheck his SATs to see if he got a 2400. If true, we would pass the right son-node of the current node and move on to check the other factors.

The disadvantage for this decision system is that the tree could be very big. The time required to build this tree may be time consuming.

This decision tree technology does not include a machine learning capability. Now, we want to add this part to the system. Assume this university only accepts 3000 students this year. The existing decision tree has rejected too many students, so we need to add some rules to the system, which would add nodes automatically based on the samples we have. If the system adopts this feature, then such a decision tree system could be considered as having machine learning abilities.

There are some sophisticated systems made for decision-tree learning including *iterative dichotomiser 3 (ID3)* [15] and *classification and regression tree (CART)* [23].

Decision trees are also associated with a very profound technology called boosting. The principle of boosting is to improve training or learning accuracy. If we have a weaker classifier meaning there will always be relatively big errors when classifying the training set (error < 50 %), can we use the classifier multiple times on part of the training set to improve the strength of the classifier? For instance, if we can only draw a vertical or horizontal line to classify a set into two categories, we would have many errors in most cases. However, when we classify the samples incorrectly, we can use the rest of the samples in the training set for another try. We can improve the accuracy a lot and this is the key idea of the boosting method.

Selecting the partial training set is not random, and we select it with almost certainty that we can improve the result. The detailed statistical analysis can be found in [15].

4.2 Neural Networks

A typical neural network (NN) consists of the input layer, hidden layers, and the output layer. Each layer contains several nodes [4, 26]. Each node in the input layer accepts a value of the feature vector. Each node of the output layer indicates the classification of the feature vector (or input vector). The layers are connected by

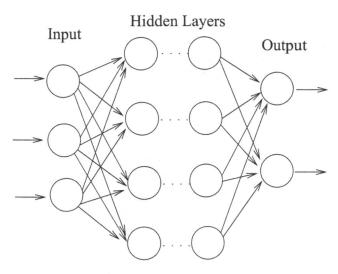

Fig. 4.1 Example of a backpropagation neural network

arcs (edges) with weights. The most popular NN is called the back-propagation neural network. It uses the least square method for the convergence of updating the weights. The detailed method can be found in [4]. See Fig. 4.1.

Neural networks is a method of statistical learning. The weights between nodes will adjust to minimize error in each iteration. The mathematical model of the simplest NN is similar to a linear transformation from the feature vector space to the categorization space that is also a set of vectors (perhaps in different dimensions). The detailed steps for training is as follows:

When we have n nodes in the input layer and m nodes in the output layer, we will have a linking line between each pair of input nodes and output nodes. When we put a weight on each linking line (arc), we will have an $n \times m$ matrix $W = [w_{ij}]$. As we have discussed, we have a training set (F, V). For the feature vector $u = (u_1, \ldots, u_n)$ that is in class i, we use $v = (v_1, \ldots, v_m)$ to represent its classification by letting $v_i = 1$ and $v_j = 0$ for $i \neq j$. In other words, for each (u, v), we require $v = (v_1, \ldots, v_m) \in V$, satisfying $v_i \in \{0, 1\}$ and $\Sigma v_i = 1$. The key idea of the learning is to modify the weights w_{ij} when inputing each sample of the training set.

The NN method is a type of statistical learning method. If hidden layers are added, more weight matrices are involved. For the purpose of effectiveness, after the matrix calculation, in NN, we usually add a nonlinear transfer function S that could be a step function. The biggest component of vector $S(u \cdot W) = z$ indicates the classification of u as the "intermediate" classification. The learning process passes $(v - z)$ back to w_{ij} to get a better result by changing the value of w_{ij}. After a certain number of iterations, w_{ij} will reach a stable value. The learning process would then stop.

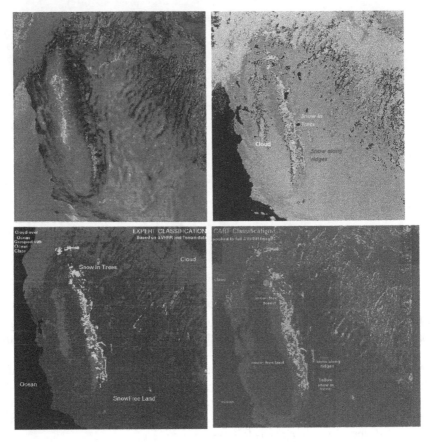

Fig. 4.2 Neural network classification: (**a**) original image, (**b**) the classification from an NN, (**c**) the classification by a decision tree, (**d**) the expert classification (Color figure online)

One of the very successful applications of NN is optical character recognition (OCR) [18]. Some researchers recognized that support vector machine (SVM) is a type of NN methods. NN has many applications. We show some results of the real data recognition of satellite image processing in Fig. 4.2 [6, 10].

4.3 Genetic Algorithms

The genetic algorithm is a specialized search technique, which is not guaranteed to arrive at a "best" solution. However, it generally can get to a good or near optimal solution in many cases [11, 12].

One example is for the travelling salesman problem, where we need to find the shortest path for a salesperson to visit each big city once in the USA and then return to the start.

The solution for this problem can be string that records the sequence of the visiting path. For instance, *ABCDEFGA* represents the salesperson leaving city *A* and going to city *B*, then *C*, and so on. He or she will finally return to *A* from city *G*.

We can calculate the distance of this path. The distance here is called a fitness function. Let us assume that *ABGFCDEA* has a shorter distance. How do we obtain the sequence or an even better solution?

A genetic algorithm consists of a function and two operations:

(1) The objective or the fitness function. It tells the genetic algorithm which paths (called chromosomes) are closer to the objective.
(2) The crossover operation: For two strings S and T, randomly select a position on a string and cut the strings into two parts $S = S_1 S_2$ and $T = T_1 T_2$, where the crossover means to have two new strings such as $S_1 T_2$ and $T_1 S_2$,
(3) The mutation operation is where we find a local element in the string and change its value. (In this example, we need to make sure that S_1 and T_1 cover the same cites.)

We return back to the example we used. We can first randomly select ten strings (starting from city A and then back to A). And calculate each fitness values for each string. We then only keep the five with the shortest distances (or the value of the fitness function). Then we randomly do crossover or mutation operations among these five strings and keep the best five. This iteration will run for a given number of times or until the results can no longer be improved. The best answer will be the solution for the genetic algorithm.

The principle of this algorithm is based on Darwins "survival of the fittest". The good "genes," such as a certain substring that would make the trip shorter, will always stay in the final string. That is the key to understanding this algorithm. For instance, after the sales person has visited Washington DC, he must visit Baltimore, or vice versa. This combination of Washington DC-Baltimore is a good gene for our problem.

4.4 Functional and Variational Learning

Variational learning is one of the most important developments in applied mathematics to image processing [7, 21]. Finding the optimization of a function in image segmentation [17, 25, 28] is one of the major developments in the application of the variational method. The principle of this method is to use harmonic functions to solve the Dirichlet problem in functional analysis [20, 21, 35].

Variational learning was first proposed by Mumford and Shah in [21]. It considers three factors in segmentation: (1) the total length of all the segments and edges, (2) the unevenness of the image without its edges, and (3) the total error between the original image and the proposed segmented images where each segment has unique or similar values in its pixels. When the three weighted factors are minimized, the resulting image is a solution of the Mumford–Shah method.

In mathematics, the variational principle came from the solution of the Dirichlet problem: given a continuous function f on boundary ∂D of D, is there a differentiable extension F of f on D? The solution, a harmonic function, is related to the minimum energy function for an integral:

$$E[f(x)] = \int_\Omega (f_x^2 + f_y^2)\mathrm{d}x\mathrm{d}y. \qquad (4.1)$$

The harmonic function is the function that satisfies the following equation.

$$\frac{\partial^2 f}{\partial x^2} + \frac{\partial^2 f}{\partial y^2} = 0 \qquad (4.2)$$

In Mumford–Shah segmentation, the learning is performed inside of each iteration. Let F be an image (in a domain D). Assume f be the model (a reconstructed image) and let B be the boundaries of each segmented component of f. The Mumford–Shah functional is defined as [21]

$$E[f, B] = \alpha \int_D (F - f)^2 dA + \beta \int_{D-B} |\nabla f|^2 dA + \gamma \int_B ds$$

where α, β, and γ are weights with $\alpha + \beta + \gamma = 1$. This formula means that the difference between the original image and fitted image should be small because the first term is $\int_D (F - f)^2 dA$. The internal variation of f should be small because of the second term (the standard segmentation has the same value in a segmented component). The total boundary length should also be small in many cases. The Mumford–Shah method has many variations now for different problems. In Fig. 4.3, a nice segmentation using a modified Mumford–Shah functional was presented by Vese and Chan in [35]. In Chap. 10, Spencer and Chen will present a more complete coverage and application in this book.

4.5   Support Vector Machine Algorithms

The SVM algorithm is another machine learning model [8]. It is an advanced method in machine learning algorithms that uses regression analysis to a certain subset of training samples. This method is to find a best separation line between two classes (a middle line to both classes) [8, 32–34].

The philosophy behind this method is to get a separation line based on some of the boundary points of two classes, and there are only two classes for a line to split. Similar to the process of Voronoi diagrams, we would like to find a best partition regarding a center point (a site). However, for a randomly shaped point that belongs

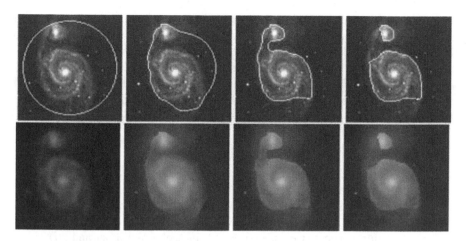

Fig. 4.3 Image segmentation using a modified Mumford–Shah functional

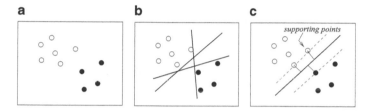

Fig. 4.4 The support vector machine: (**a**) a training set with two classes, (**b**) three possible classifiers shown as three lines, and (**c**) the best separation line supported by three points

to two categories, having a center point for a class or category does not make sense. Using the boundary of edges will give the most meaning in many cases. For instance, the boundary of two countries can be separated by some small towns near the border. How do we draw an actual borderline? We can find a line that is equidistant from the towns in both countries. We call the towns supporting points.

This is a good philosophy in practical learning. The rest of this method is easy to understand, albeit technically difficult in practice. For instance, we can draw multiple lines along the border, such as polylines linked together. We can even fit a curve to approximate the polylines. To summarize, k-mean and k-NN use centers for classification, while SVM uses boundaries for classification.

In Fig. 4.4, for an original data set (a), we can find three reasonable lines to separate them (b). However, line a will give the largest gap. So a will be selected by SVM. The data points A, B, and C will be called the supporting vectors.

Mathematically, let $X = \{x_1, x_2, \cdots, x_n\}$ be a training set where x_i is an m-dimensional feature vector. These points (training samples) belong to class ω_1 and ω_2; they are separable by a line in 2D or a hyperplane in high dimensional space. A line or hyperplane is in the following form:

$$f(x) = Ax + c_0 = 0$$

A is a matrix for rotation and c_0 is a constant (may be related to the distance from ω_1 or ω_2). If $f(x)$ splits these two classes completely, then when we put x_i into the equation, we will get a positive number for ω_1 and a negative number for ω_2, and vice versa. Now, our objective is to get the equation to be equidistant from ω_1 and ω_2. At the same time, we want the distance to be the largest. Based on analytical geometry, the distance from x to $f(x)$ is $|f(x)/||A|||$. We would like to indicate that the closest point in ω_1 is $+1$ and the closest point in ω_2 is -1 for simplicity. Thus,

$$\begin{cases} Ax + c_0 \geq 1, x \in \omega_1 \\ Ax + c_0 \leq -1, x \in \omega_2 \end{cases}$$

Then, the margin of both sizes of $f(x)$ will be $2/||A||$. To maximize the margin is to minimize $||A||$ or $||A||^2$ for smooth derivative calculations.

For calculation purpose, we use $c(i) = +1$ for each point in w_1 and $c(i) = -1$ for each point in w_2. Now, we can put our optimization as

$$\begin{cases} \min E(A) = 1/2||A||^2 \\ c(i)Ax_i + c_0 \geq 1, i = 1, 2, \cdots, n. \end{cases} \tag{4.3}$$

Using the Lagrange multiplier, we can solve the above optimization problem. For the detailed method, see [29, 32].

When we treat a massive data set, we can split the classified data samples into different machines in a cloud. When we get the first line from each machine, we can link them in a way that creates a continuous joint point on the border lines. We can even modify the neighborhood points a bit to make them "smooth" looking at the joint point.

Before SVM can be applied, we usually need a classification method for finding $w1$ and $w2$. We can use the k-mean or k-NN method for the initial classification. It is also possible to only use the modified SVM for direct classification and finding the optimum margin. The SVM method can be very complicated when a kernel is involved to predict curves as separation.

4.6 Computational Learning Theory

Computational Learning Theory focuses on learning algorithms for the polynomial time learnable functions. There are two major methods in computational learning: (1) Probably approximately correct (PAC) learning, and (2) Vapnik–Chervonenkis theory. These two methods are somewhat equivalent.

PAC learning theory was proposed by Valiant [30, 31]. We will give a brief introduction here. Given a set of training samples T, $(x_i, f(x_i))$, we would like to guess the function $f(x)$, where x is a member of a domain X. The guessed function is called a hypotheses $h(x)$. Nilsson gave a very nice lecture on this method in [22].

The purpose is to make $h(x) = f(x)$ for a restricted set having a high probability of being correct (or approximately correct, in other words, having a small probability for error). Such a restricted set can also be a testing set that has the same distribution of f on X.

Let H be a set of hypotheses, $|T| = m$, and $|H| = N$.

Theorem 4.1. *If T contains $m > \frac{1}{\epsilon}(\ln N + \ln \frac{1}{\delta})$ independent training samples, the probability that there exists a hypothesis $h \in H$ that is consistent with f on these samples and has error greater than ϵ is at most δ.*

A class, F, is (polynomially) PAC learnable (when H is provided, and $F \subset H$) if there exists a polynomial-time learning algorithm to learn each function f in F.

Here, the term "polynomial" is about m, the size of the input in this instance ($x \in X$ and x is coding for the problem), where $\frac{1}{\epsilon}$ and $\frac{1}{\delta}$.

The Vapnik–Chervonenkis (VC) theory is profound [33]. We only introduce the concept of the VC-dimension here. For any set S, the VC-dimension is the smallest number of functions hypothesis (lines or circles, definable in a finite way) that can split S into any two disjoint sets. Let H be a set of hypotheses. We say that S is shattered by H if for any bi-partition of S (meaning partitioned S into two classes), there is an $h \in H$ where h is consistent with this partition. For instance, S contains three points in 2D. For any type of classification of S into two classes, we can use one line to split S to match the classification. However, if S contains four points, then one line is not enough in some settings.

Therefore, the VC-dimension for S with four points in 2D is 2. In 1987, Blumer *et al.* proved an important theorem, which we present below [5].

Theorem 4.2. *A hypothesis space H is (polynomial) PAC learnable if and only if it has finite VC-dimension.*

Let us use an example to explain PAC learning. We try to use the letter "B" as a learnable example. "B" is in a binary image form that is coded as a binary string. The image can be small or big, and the length of the string is called the size of the input. The training set consists of many example points where some are marked "+1" if they are positive and "−1" if they are negative. H is the set of the hypotheses. $h \in H$ can be as follows: "letter B has two holes," "there are at least ten elements in the image array as 1," "there is a line segment that is almost straight," etc.

In order to make this learning system work, our training examples must exceed a certain number. F can be a subset of H for simplicity.

4.7   Remarks: Statistical Learning Algorithms and BigData

Comparing to many learning theories such as rule-based learning, decision-trees, and genetic algorithms, statistical learning is still the most popular. Most of the statistical learning algorithms use numerical learning that mainly refers to a regression method [1–3, 16]. It usually uses the least square technology. Most readers with mathematical backgrounds can easily understand this method. However, for nonlinear models, the regression is much more complicated. Machine learning and artificial intelligence methods can be used to partition or segment the original models by treating each segment as a linear model. We have introduced least square regression in Chap. 2. In Chap. 9, Huang et al will focus on methods for financial data analysis.

For unsupervised learning k-mean may be still best. However, for massive data sets or BigData, k-mean might not be very practical due to the extensive iterations. However, k-NN for supervised learning can be easily turned into a parallel algorithm. For instance, we can distribute the calculation to each class in different machines. It is a good idea to investigate the k-mean algorithm in a MapReduce programming style [9, 13, 36], and some research have been done on k-mean++ [1, 37].

For learning methods: Bayesian Networks and the Mean Shift method, We would like to mention two other Bayesian networks are also called belief networks. They are used for inference and reasoning. Bayesian networks use the directed graph to represent these variables (as vertices) and the causal connection (as arcs/edges). It is somewhat similar to λ-connectedness. The difference is that the weight on arcs is followed by probabilistic distributions.

Let $U = \{X_1, \ldots, X_n\}$ be a set of random variables. For example, if there is an arc from vertex $X1$ to vertex $X2$, we know that $X1$ causes $X2$ (see Fig. 4.5). Each node associates with a table that indicates the conditional probability, called conditional probability (distribution) table [24]. The summation of all outer probabilities is equivalent to the probability on the node. The solving path is to backpropagate the source. This is the reason why a neural network can be viewed as a type of Bayesian networks.

Fig. 4.5 Example for Bayesian networks [7]

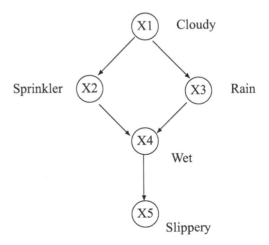

Figure 4.5. shows if the grass is slippery, then it is wet. This is either caused by rain or the sprinkler which was turned on. The Bayesian network can tell us which one is most likely based on the joint distribution formula.

$$P(x_1, \ldots, x_n) = \Pi_i P(x_i | Parent(i)) \qquad (4.4)$$

where x_i is the value of X_i, and $Parent(i)$ is all values of parent notes of X_i.

Therefore, the weight of each arc will become reality after an assumption is made or a real event that has appeared. The path of propagation reasoning is then similar to a λ-connected path we discussed in Chap. 3.

The method of Bayesian networks was recently used with the MapReduce framework and Hadoop clusters for parameter learning [2, 3]. Basak et al have tested the expectation maximization (EM) learning model for both sequential iteration (SEM) and MapReduce-framework iteration (MREM). SEM on the ADAPT T2 network used 2 h 30 min per iteration for 100,000 data records. For the same task, MREM on an Amazon EC2 cluster used only 15 min for an iteration. That is about 10 times faster using Hadoop clusters. (Hadoop is a type of implementation of the MapReduce framework.)

The mean shift method tracks a set of data points based on the density function. This method is very popular in object recognition and tracking in video processing. We discuss the details in Chap. 5. In this book, we have a chapter for statistical learning, specifically the Monte Carlo method (see Chap. 7).

The relations among data mining, machine learning, and statistics can be found in [14, 15, 19].

Acknowledgements Many thanks to Professor Nils. J. Nilsson for agreeing the author to use his unpublished notes.

References

1. B. Bahmani, B. Moseley, A. Vattani, R. Kumar, S. Vassilvitskii, Scalable k-means++, in Proc. VLDB Endowment **5**(7), 622–633 (2012)
2. A. Basak, I. Brinster, X. Ma, O. Mengshoel, Accelerating Bayesian network parameter learning using Hadoop and MapReduce, in *Proceedings of the 1st International Workshop on Big Data, Streams and Heterogeneous Source Mining: Algorithms, Systems, Programming Models and Applications , BigMine 12* (ACM, New York, NY, 2012), pp. 101–108
3. A. Basak, I. Brinster, O.J. Mengshoel, MapReduce for Bayesian Network Parameter Learning using the EM Algorithm, in *Proceedings of Big Learning: Algorithms, Systems, and Tools*, 2012. http://www.repository.cmu.edu/silicon_valley/151/
4. C.M. Bishop, *Neural Networks for Pattern Recognition* (Clarendon Press, Oxford, 1995)
5. A. Blumer, A. Ehrenfeucht, D. Haussler, M. K. Warmuth, Learnability and the Vapnik–Chervonenkis dimension. J. ACM **36**(4), 929–965 (1989)
6. L. Chen, Possibility functions based fuzzy neural networks and their applications to image processing, MS Thesis of Computer Science, Utah State University, Logan, USA, 1995.
7. L.M. Chen, Digital and Discrete Geometry: Theory and Algorithms, NY Springer, 2014.
8. C. Cortes, V. Vapnik, Support-vector networks. Mach. Learn. **20**(3), 273–297 (1995)

9. J. Dean, S. Ghemawat, MapReduce: simplified data processing on large clusters. Commun. ACM **51**(1), 107–113 (2008)

10. C. Feng, A. Sutherland, R. King, S. Muggleton, R. Henery, Comparison of machine learning classifiers to statistics and neural networks, in *AI &Stats-93*, pp. 41–52, 1993.

11. A.A. Freitas, *Data Mining and Knowledge Discovery with Evolutionary Algorithms*, 2nd edn. (Springer, Berlin, 2013)

12. D.E. Goldberg, *Genetic Algorithms in Search, Optimisation and Machine Learning* (Addison-Wesley Publishing, Boston, 1998)

13. Hadoop: open source implementation of MapReduce. http://www.lucene.apache.org/hadoop/, 2014

14. J. Han, M. Kamber, *Data Mining: Concepts and Techniques* (Morgan Kaufmann, San Francisco, 2001)

15. T. Hastie, R. Tibshirani, J. Friedman, *The Elements of Statistical Learning: Data Mining, Inference, and Prediction*, 2nd edn. (Springer, New York, 2009)

16. A.K. Jain, R.P.W. Duin, J. Mao, Statistical pattern recognition: a review. IEEE Trans. Pattern Anal. Mach. Intell. **22**(1), 4–37 (2000)

17. R. Klette, *A Concise Computer Vision* (Springer, Berlin, 2012)

18. Y. LeCun, B. Boser, J.S. Denker, D. Henderson, R.E. Howard, W. Hubbard, L.D. Jackel, Backpropagation applied to handwritten zip code recognition. Neural Comput. **1**(4), 541–551 (Winter 1989)

19. D. Michie, D.J. Spiegelhalter, C.C. Taylor, *Machine Learning: Neural and Statistical Classification* (Ellis Horwood Limited, Cambridge, 1994)

20. J.M. Morel, S. Solimini, *Variational Methods in Image Segmentation* (Birkhauser, Basel, 1994)

21. D. Mumford, J. Shah, Optimal approximation by piecewise smooth functions and associated variational problems. Commun. Pure Appl. Math. **42**, 577–685 (1989)

22. N.J. Nilsson, *Introduction to Machine Learning*. Draft of Incomplete Notes, 1998. Available Online at: http://www.ai.stanford.edu/~nilsson/mlbook.html

23. R. Nisbet, *Handbook of Statistical Analysis and Data Mining Applications*, NY, Academic Press (2009)

24. J. Pearl, *Probabilistic Reasoning in Intelligent Systems: Networks of Plausible Inference* (Morgan Kaufmann, San Mateo, CA, 1988)

25. A. Rosenfeld, A.C. Kak, *Digital Picture Processing*, 2nd edn. (Academic, New York, 1982)

26. D.E. Rumelhart, G.E. Hinton, R.J. Williams, Learning representations by back-propagating errors. Nature **323**(6088), 533–536 (1986)

27. S. Russell , P. Norvig, *Artificial Intelligence: A Modern Approach*, 3rd edn. (Pearson, Boston, 2009)

28. H. Samet, *The Design and Analysis of Spatial Data Structures* (Addison Wesley, Reading, MA, 1990)

29. S. Theodoridis, K. Koutroumbas, *Pattern Recognition* (Academic, Orlando, FL, 2003)

30. L. Valiant, A theory of the learnable. Commun. ACM **27**, 1134–1142 (1984)

31. L. Valiant, *Probably Approximately Correct: Nature's Algorithms for Learning and Prospering in a Complex World*, reprint edn. (Basic Books Publisher, New York, 2014)

32. V.N. Vapnik, *The Nature of Statistical Learning Theory* (Springer, Berlin, 1995)

33. V. Vapnik, A. Chervonenkis, On the uniform convergence of relative frequencies of events to their probabilities. Theory Probab. Appl. **16**(2), 264–280 (1971)

34. V.N. Vapnik, S. Kotz, *Estimation of Dependencies Based on Empirical Data* (Springer, Berlin, 2006)

35. L. Vese, T. Chan, A multiphase level set framework for image segmentation using the Mumford and Shah model. Int. J. Comput. Vis. **50**(3), 271–293 (2002)

36. T. White, Hadoop: The definitive guide, 2012

37. W. Zhao, H. Ma, Q. He, Parallel k-means clustering based on MapReduce, in *CloudCom*, 2009, pp. 674–679

Chapter 5
Images, Videos, and BigData

Li M. Chen

Abstract Images and videos are really BigData, even before the concept of BigData was initiated or created. For many companies, image related data occupies 80 % of their storage. Therefore, data processing related to images is an essential topic in data science. The tasks concerning images and videos are mainly object search, recognition, and tracking. Current and future applications of images and videos include security and surveillance, medical imaging, traffic monitoring, industrial measurements, document recognition, automated driving, and more. In this chapter, we focus on massive image data processing and computer vision. We will still focus on machine learning algorithms. Images and video always require most of the storage space and by having applications over the Internet, we can say that image related problems are always BigData problems. Even with other applications, we still need to consider massive data processing. For instance, automated driving is a challenge to data science.

In BigData related image processing, we will discuss the following topics in this chapter: (1) An overview of image and video segmentation, (2) Data storage and fast image segmentation, (3) Feature extraction, (4) Learning and training, and (5) Classification and decision making.

5.1 Images and Videos in BigData Times

Today, BigData related images and videos appear everywhere such as security, surveillance, object tracking, facial recognition, document recognition, automated driving, medical imaging, and meteorological and satellite images. Along with the rapid growth of multiple structured or even unstructured data, in the future, image processing must be able to assimilate and interpret images and videos in the same way we are able to analyze text and numbers now [39]. Image and video processing is very broad. In this chapter, we focus on the data preprocessing and BigData related parts such as image segmentation and video tracking. The purpose of image analysis and recognition is to make a final decision that is related to prediction and

L.M. Chen (✉)
The University of the District of Columbia, Washington, DC, USA
e-mail: lchen@udc.edu

© Springer International Publishing Switzerland 2015
L.M. Chen et al., *Mathematical Problems in Data Science*,
DOI 10.1007/978-3-319-25127-1_5

interpretation (prescription). Prior to decision making, we will need to collect data (BigData) to prepare an analytical data, select and use predictive models, and make predictions and interpretations [39].

Images and videos in BigData times require that we use fast algorithms in many cases. We must be able to use multiple machines on cloud computers, and we must be able to adapt the different images or video formats over the Internet. On the other hand, moving object tracking in videos, called video tracking, seeks to locate a moving object or multiple objects in time stamped videos. It is probably one of the most widely used techniques in image processing and computer vision today. Just to name a few areas where such technology would be useful: human–computer interaction and video editing, security and surveillance, video compression and communication, traffic control, and medical imaging.

Some of the learning algorithms are not suitable for these purposes when learning requires extensive time. If the data set is too large, the algorithm for some statistical methods cannot be directly applied since some matrix operations require $O(n^2)$ scale level space in the memory of a single machine. For instance, finding eigenvalues and eigenvectors of the Laplacian matrix has many applications from the finite difference method to the graph-cut method. If the matrix is too big, then this type of calculation may need to be changed to an approximate but fast algorithm [9].

Even though there is a good amount of research in statistics and pattern recognition, algorithmic learning over the Internet, in a parallel or distributed manner, potentially using new mathematics and statistics, is the biggest challenge today in data science. The main concern is with the linear $O(n)$ time, sub-linear $(log(n), \sqrt{(n)})$ time, and log-liner $(O(n \log n))$ time complexities. Video data is a sequence of images, but it is usually in a compressed form. The process of using video data may require the restoration of the data.

Some real-world examples and problems of massive image data include the following:

(1) Image forensics. A face recognition system is essential for this field to locate a criminal record for a person whose face was captured by a camera. The system needs to export this image from a recorded video.
(2) Security and surveillance. We need to use motion recognition in videos. Recognized certain human or moving activities.
(3) Medical imaging. Data statistics is needed to list different features in the collection of diseases. This would require multiple formats and resources in different hospitals and data locations.
(4) Automated vehicle driving. For instance, the company ImageVision announced that it can detect changes in video, frame by frame, by classifying and categorizing actions and motions.
(5) Text recognition and data extraction. This can be used to transfer an old newspaper image into text. This could also be used in vehicle license plate recognition in real time.

Surveillance and image forensics usually require a human to make a selection. After that, we need to put it into an image analysis system for image cleaning and enhancement since video recordings usually contain a lot of noise. After a relatively clean image is generated, we need to upload the image into various databases and Internet image collections. This task is time consuming and requires cloud computing techniques. When we require Hadoop or SPARK coding for a fast resolution, we need to be familiar with the programming practice [40]. There are also some existing companies that can help in the search and analysis such as www.ImageVision.com. For instance, the ImageVision software system developed for facial recognition is an automated, real-time image and video manipulation detection program with an easy-to-use API (the application programming interface). There are some manipulation examples including photo retouching and adding or removing content, which researchers can also use.

5.2 Concepts of Image Processing: Filtering, Segmentation, and Recognition

From a technical point of view, image and video processing, including computer vision, can be divided into five main tasks [16, 20]: (1) image filtering that makes clear or clean images for preprocessing, (2) image segmentation that partitions the image into meaningful parts, (3) feature selection, (4) recognition that includes prediction, and (5) interpretation that includes prescription.

The first three tasks make data for the next two tasks. Image filtering is considered to use sophisticated technology already, and feature selection is relatively simple. Feature selection forms feature vectors for a local point, a small collection of pixels (a square), or a component in an image. Therefore image segmentation is the main focus in this chapter. For image recognition, people use methods with similar techniques to classification and machine learning, which we discussed in Chaps. 2 and 4.

In other words, methods in pattern recognition, statistical reasoning, and machine learning can be used. We have discussed many classification methods in the previous chapter and most other technologies introduced in that chapter can be used here.

In this chapter, we will focus on image segmentation and moving object search and tracking in videos.

Image processing is very broad. The main tasks related to mathematics are filtering, segmentation, and recognition. These are some of the topics in mathematical image processing specifically discussed in this book

5.2.1 Image Filtering

The digital form of images is either a raster image or a vector image. A raster image is saved in picture elements (pixels), each of which represents a small unit area of a continuous picture. Pixels are arranged as a sequence in computer memory and have numerical values for intensity (brightness) of gray scale or color. Our cell phone pictures are stored in this format. Vector images are formed by a set of polygons with an intensity or color value. The printed characters (fonts) are usually stored in vector image forms. Thus, the size change of the character does not generate jagged edges.

There are two basic techniques for image filtering. One is to average the surrounding pixel values with a center pixel. If there is a noise pixel, then we can remove it or give it a more reasonable value relative to that of the center pixel. This is called a low frequency pass. The value we assign is not very different from that of adjacent pixels. Another way is to use the center pixel to subtract adjacent pixels. This is called a high pass meaning that the center pixel will pass if it has a large difference from its neighbors. It is used to detect the edges of objects in an image. The mathematical foundations of image filtering come from the Fourier transform, which we describe below [16].

Let $f(t)$ be a periodic wave function that has an amplitude (intensity) at time t. Also, $e^{jx} = cos(x) + jsin(x)$ is a complex number where $j = \sqrt{-1}$ is the imaginary unit.

The Fourier transform of $f(t)$ represents the function on frequency domain U.

$$F(u) = \int_{(-\infty)}^{\infty} f(x)e^{(-j2\pi ut)dt}. \tag{5.1}$$

u represents frequency. The inverse Fourier transform is the following:

$$f(t) = F^{-1}(t) = \int_{(-\infty)}^{\infty} F(u)e^{(j2\pi tu)du}.$$

Scientists use part of the frequency such as $u \in [-B, B]$ to reconstruct $f(t)$. When we think about the original $f(t)$, it contains much noise beyond $[-B, B]$ in its frequency domain.

To use the Fourier transform to filter an image and remove noise, we need to do a convolution as defined below:

$$(f * g)(t) = \int_{(-\infty)}^{\infty} f(\tau)g(t - \tau)d\tau.$$

Understanding this formula in signal processing is not difficult.

The convolution is equivalent to the average of the surrounding pixels. This process will overcome some noise pixels (low frequency signals will pass easily).

When we want to enhance edge information in objects in the image, we want to make a high frequency signal filter. We subtract the value from a neighbor pixel (using the derivatives) to make such a filter.

5.2.2   Segmentation

After an image is filtered, we would like to partition the image into pieces where each piece represents an object. The piece is also called a segment or component. We have introduced the concept of image segmentation in Chap. 3. Image segmentation is the basic approach in image processing and computer vision [28]. It is used to locate special regions and then extract information from them.

Image segmentation is an essential procedure for image preprocessing, object detection and extraction, and object tracking.

Even though there is no unified theory for image segmentation, some practical methods have been studied over the years such as thresholding, edge based segmentation, region growing, clustering (unsupervised classification), and split-and-merge segmentation to name a few.

The methods that use profound mathematics are variational image segmentation and graph-cut segmentation. We have introduced the principle of variational methods in Chap. 4 and will discuss graph-cut image segmentation in this chapter.

For thresholding image segmentation, the key is to find the best threshold to cut an image into a binary image. We will discuss two popular methods in this chapter: the maximum entropy method and Otsu's minimum variance method [16, 27].

Edge based segmentation is to enhance the edge of each component. We can use filtering techniques to enhance the edge information, and then use a search algorithm such as chain-code to extract the boundary of a component.

5.2.3   Image Recognition

In general, image recognition is part of pattern recognition. For instance, for a location $p = (i, j)$ in an image, we can use 3×3 pixel arrays centered at p as a vector to do classification. This vector has nine elements. Most methods introduced in Chaps. 2 and 4 can be applied to this type of image recognition. Decision trees, k-NN and neural networks are good recognition methods.

There is a so-called object recognition method in image processing. This type of recognition needs special treatment other than a simple vectorization [16, 20]. For object recognition, the key is to get a good image segmentation, not only to obtain pixel values. After a component is identified by image segmentation, we can recognize it by collecting the characteristics of the components. The characteristics can be represented by a feature vector as well. The difficult part is how we select the features. For example, in character recognition, we can use topological holes to distinguish "A" and "L." Other features include projection, histograms, and orientation.

In the last section of this chapter, we will discuss moving object recognition that uses the mean-shift method and color histograms. There are some techniques where image segmentation and object recognition will work together to obtain better results in image segmentation, object recognition, or both.

Therefore, image segmentation is an absolute key to image processing and computer vision. As we discussed in Chap. 4, variational methods have made significant contributions to image segmentation in the application of mathematics.

5.3 The Five Philosophies of Image Segmentation

There is no unified theory for image segmentation. Different problems require different segmentation methods. Some practical methods have been studied over the years such as thresholding, edge based segmentation, region growing, clustering (unsupervised classification, e.g. k-mean), split-and-merge segmentation, and the graph-theoretical based method. These segmentation algorithms have been developed for solving different problems [8, 28]. However, they are all based on one or more of the five philosophies listed below:

(1) A segment is a class or cluster, so we can use a classification or clustering method to segment the image. Classification methods usually do not need to use the location position information. Clustering for unsupervised classification technology can perform better to find an object for sampled points within the subset of data frames. Typical techniques include k-mean. This method classifies the pixels into different groups in order to minimize the total "errors," where the "error" is the distance from the pixel value to the center of its own group (the average value).

(2) Segmentation is to separate foreground and background. It is to find a clip-level to make a binary image. If an object or region can be identified by absolute intensity (the pixel value), we usually use threshold segmentation. In other words, an object will be recognized as a geometrically connected region whose values/intensities are between a certain high-limit and low-limit. We usually assume that the high limit is the highest value of the image. Therefore, in practice, we only need to determine the low-limit. Maximum entropy and minimum variance (also called Otsu's method) are two of the most popular methods for determining the best threshold for a single image [27, 34].

Multilevel thresholding is similar to threshold segmentation and uses the same philosophy, but multiple thresholds are produced at once. It requires an extremely high time cost for computation [17, 19, 28]. We will discuss these two methods in detail in Sect. 5.4.

(3) A segment is a homogeneous region or a "smoothly connected" region. In a region where intensity changes smoothly or gradually, the region is viewed as a segment. A popular segmentation method is called mean-based region growing segmentation in [16]. A pixel will be included in a region if the updated region is homogeneous, meaning that the difference between pixel intensity and the mean of the region is limited by ϵ, a small real number.

The λ-connected segmentation introduced in Chap. 2 follows the same philosophy. This is to link all pixels that have similar intensity [10, 38].

The big-pixel technology can be viewed as an extension of this method. In this method, an image is first partitioned into some small regions called big pixels. Each small region is absolutely part of an object. Then, we use a graph to link all big pixels and then use the mean-based region growing method or λ-connectedness to make the final segmentation. The Canon company has obtained a considerably important patent in the USA where the patent document shows the relationship among those methods. [41]

(4) Split-and-merge segmentation uses quadtree to determine the order in which pixel(s) should be treated or computed [18, 29, 36]. It is an algorithmic way to find an object or force a merge order. This is because the method is based on the mean of the merged region. It does not guarantee a transitive relation. Again, the mean-based segmentation is not an equivalence relation. This method splits an image into four sections and checks if each part is homogenous. The homogenous segments are then merged together.

If the segments are not homogenous, the splitting process is repeated. This process is also called quadtree segmentation. The method is more accurate for some complex images, but it costs more time to segment an image.

The time complexity of this method (process) is $O(nlogn)$. This was proved by Chen in 1991 [4, 31].

(5) A segment is surrounded by one or several closed boundary curves (edges). If we can detect and track the edges of a region or segment, we can determine the location and outline of the segment. The fifth philosophy is edge detection. Finding low or high frequency pixels is very common in edge detection. However, not all edge-detection methods can be used in image segmentation since enhancing an edge is not the primary purpose of image segmentation. The purpose of image segmentation is to find components.

In addition, we usually use the so-called chain-code method to track the edge of an object. As we discussed in Sect. 5.2 of this chapter, we may use an edge enhancement technique to increase the signal on the edges. The chain-code algorithm follows the edge of a component by always turning clockwise or counterclockwise while finding the next pixel on the edge.

The number of edges on the boundary should be relatively small. Otherwise, the extraction of the closed curves will be a major problem. The Mumford–Shah method, which uses the variational principle [25], is an important development in image segmentation. In their method, the total length of the boundaries of the segments will be specifically considered. See details in Chap. 4.

5.4   Image Segmentation Methods

In this section, we first introduce two advanced segmentation methods: the maximum entropy method and Otsu's minimum variance method. Then we discuss the combination of λ-connectedness with these two methods. We will also discuss the possible connections between these methods to the Mumford–Shah method.

5.4.1   The Maximum Entropy Method and the Minimum Variance Method

The maximum entropy method was first proposed by Kapur, Sahoo, and Wong [19]. It is based on the maximization of the inner entropy in both the foreground and background. The purpose of finding the best threshold is to make both objects in the foreground and background, respectively, as smooth as possible [19, 28, 34].

If F and B are in the foreground and background classes, respectively, the maximum entropy can be calculated as follows:

$$H_F(t) = -\Sigma_{i=0}^{t} \frac{p_i}{p(F)} \ln \frac{p_i}{p(F)}$$

$$H_B(t) = -\Sigma_{i=t+1}^{255} \frac{p_i}{p(B)} \ln \frac{p_i}{p(B)}$$

where p_i can be viewed as the number of pixels whose value is i; $p(B)$ is the number of pixels in the background, and $p(F)$ is the number of pixels in the foreground. The maximum entropy is to find the threshold value t that maximizes $H_F(t) + H_B(t)$ See, Fig. 5.1.

Minimum variance segmentation was first studied by Otsu in image segmentation [8, 27]. Otsu's segmentation was the first global optimization solution for image segmentation. It is used to clip the image into two parts: the object and the background.

Assume that $\sigma^2(W)$, $\sigma^2(B)$, and $\sigma^2(T)$ represent the within-class variance, between-class variance, and the total variance, respectively. The optimum threshold will be determined by maximizing one of the following criteria with respect to threshold t [8, 27]:

$$\frac{\sigma^2(B)}{\sigma^2(W)}, \frac{\sigma^2(B)}{\sigma^2(T)}, \frac{\sigma^2(T)}{\sigma^2(W)}$$

σ is the standard deviation. Since $\sigma^2(T)$ is constant for a certain image, this segmentation process makes between-class variance large and within-class variance small. Therefore, our task is to make the within-class variance as small as possible.

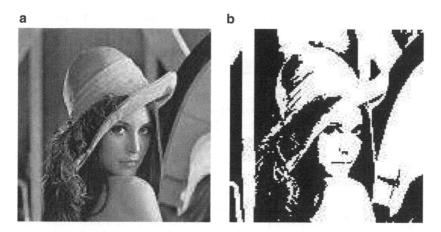

Fig. 5.1 Image Segmentation for testing image "Lena": (**a**) The original image, (**b**) Result of maximum entropy segmentation

5.4.2 Learning in λ-Connectedness with Maximum Entropy

Chen proposed a method that uses the maximum entropy method to determine the λ value [8]. It can be called maximum entropy connectedness determination. We know that the maximum entropy is to find the threshold value t that maximizes $H_F(t) + H_B(t)$.

Such an idea can be used for λ-connected segmentation. However, the total inner entropy for the image is the combination of the entropy for each segment (λ-connected component), not for the threshold clipped foreground/background. This is because in λ-connected segmentation there is no specific background [6]. Each λ-connected segment can be viewed as a foreground, and the rest may be viewed as the background. It is different from the original maximum entropy where the range of pixel values determines the inclusion of pixels. Therefore, we need to summarize all inner entropies in all segments.

$$H(\lambda) = \Sigma(\text{inner entropy of each } \lambda\text{-connected component}) \qquad (5.2)$$

We select the λ such that $H(\lambda)$ will be maximized. We call this λ-value the maximum entropy connectedness. This unique value is a new measure for images.

Since maximum entropy means the minimum amount of information or minimum variation, we want minimum changes inside each segment. This matches the philosophy of the original maximum entropy method. In other words, the λ-connected maximum entropy has a better meaning in some applications. We use the λ_e such that

$$H(\lambda_e) = \max\{H(x) | x \in [0, 1]\}.$$

Assume there are m λ-components. Define the inner entropy of each λ-component S_i:

$$H_i(\lambda) = \Sigma_{k=0}^{255} - \frac{Histogram[k]}{n} \log \frac{Histogram[k]}{n}$$

where n is the number of points in component S_i. $Histogram[k]$ is the number of pixels whose values are k in the segment. Thus,

$$H(\lambda) = \Sigma_{i=1}^{m} H_i(\lambda)) \tag{5.3}$$

The maximum entropy connectedness can be viewed as a measure of a special connectivity of the image. If the λ value is calculated in the above formula for an image that maximizes $H(\lambda)$, we say that the image has maximum entropy connectedness λ, denoted by λ_e. Some experimental results with λ_e are shown as follows.

In [7], we proposed a golden cut method for finding the λ-value for bone density connectedness calculation. We have obtained $\lambda=0.96$, 0.97 for a bone image (the size of the picture is different from the one used in this book). For a similar image, using the maximum entropy connectedness presented in this section, we get $\lambda_e=0.95$. The result is quite reasonable. See Fig. 5.2.

Using maximum entropy is a type of philosophical change. In fact, we can consider other formulas. For example, we can select another way of calculating the entropy of one segment.

We proposed a different formula in [8]. We can calculate the inner entropy of a component, then treat the rest of the image as the background for the component. The total entropy generated by this segment is the summation of both. We can apply this process to all components/segments while segmenting.

Let I be an image, $C_i(\lambda) = I - S_i(\lambda)$ is the complement of component $S_i(\lambda)$

$$H(C_i(\lambda)) = \{\text{Entropy for the set } C_i(\lambda)\}$$

We can use the following formula for the basis of optimization.

$$H(\lambda) = H(S_i(\lambda)) + H(C_i(\lambda)).$$

The outer (background) entropy is the total.

$$H(outer) = \Sigma H(C_i)$$

In $H(outer)$, a pixel is calculated multiple times.

We may need to use the average $H(outer)/m$ where m is the number of segments. The relationship between this formula and the formula we used in the previous subsection is also interesting.

Fig. 5.2 Bone density image
segmentation: (**a**) The
original image, (**b**) λ_e=0.95

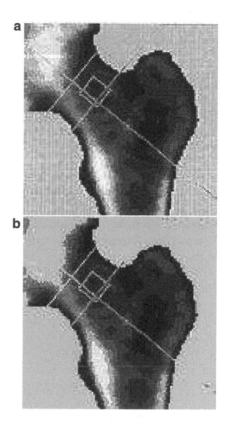

Furthermore, we should consider the following general model.

$$H_optimal = a \cdot H(inner) + b \cdot H(outer)$$

where a and b can be constants or functions of segmentation.

5.4.3 λ-Connected Segmentation and the Minimum Variance Method

It is also possible to use the minimum variance-based method for finding the best λ value in λ-connected segmentation [8].

The original design of Otsu's method is not able to be implemented directly for λ-connected segmentation, which is similar to the case of maximum entropy. This is because there were only two categories, the foreground (objects) and the background. For the first and second criteria, in λ-connected segmentation, there are many components and it would be very hard to find between-class variance.

We could consider the total between-class variance by considering every pair of components, or we could consider the between-class variance for the components that are neighbors. The third criterion seems likely to be valid. However, when we only consider the variance within a connected component, what would happen is $\sigma^2(W) = 0$ if $\lambda = 1$. The value of $\frac{\sigma^2(T)}{\sigma^2(W)}$ will be infinite and will always have the greatest value.

In order to make use of Otsu's philosophy, we modify the original formula by adding a term that is the number of segments or components, M. We try to minimize the following formula:

$$H(\lambda) = \Sigma(\text{inner variance of each} \lambda - component) + c \cdot M \qquad (5.4)$$

where c is a constant. We could let $c = 1$. The calculation of the inner variance of a λ-component is to compute the variance (square of standard deviation) of the pixels in the component.

The following formula is to find the minimum average variance (for each component).

$$H(\lambda) = \Sigma(\text{inner variance of each} \lambda - component)/M \qquad (5.5)$$

We want to find λ_v such that

$$H(\lambda_v) = \min\{H(\lambda) | \lambda \in [0, 1]\} \qquad (5.6)$$

This strategy works for meteorological data. The experimental results show that the method is promising. For the other two kinds of images tested in maximum entropy connectedness, "Lena" and the bone image, we still need to find an appropriate method under the minimum variance philosophy.

The following images show the process on the same picture with a preprocessing threshold cut using the maximum entropy cut or 45 % peak cut. Then, we perform the automated process of finding the λ-value. Figure 5.3 shows that we arrived at λ_v=0.97 using the method of minimum variance connectedness determination described in this subsection. Without smoothing the original image, we pre-cut the image using the maximum entropy threshold. The result is not what we expected.

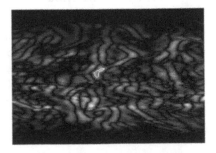

Fig. 5.3 Minimum variance connectedness determination without smoothing

Fig. 5.4 Minimum variance
connectedness determination
with smoothing

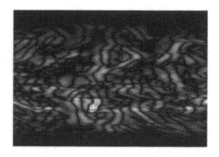

When we smoothed the image, we got λ_v=0.90, and the result turned out to be
correct. See Fig. 5.4.

5.4.4 λ-Connectedness and the Mumford-Shah Method

How do we use Mumford–Shah's idea to find the optimal segmentation? We can
define L as the total length of the edge of all segments.

$$H(\lambda) = \alpha \cdot \Sigma(\text{inner variance of each } \lambda\text{-component}) + \beta \cdot L \qquad (5.7)$$

where α and β are constants. More generally, we can do the normal λ-connected
fitting [4, 9] on each λ-connected component. The total variance (or standard
deviation) of the (normal λ-connected) fitted image is denoted by V. L is still the
total length of the edges of the segments (components), and D is the difference
between the fitted image and the original image. Using the Mumford–Shah method,
we can minimize the following equation to get the λ_p.

$$H(\lambda) = \alpha \cdot (V) + \beta \cdot (L) + \gamma \cdot (D) \qquad (5.8)$$

$$H(\lambda_p) = \min\{H(\lambda)|\lambda \in [0, 1]\} \qquad (5.9)$$

Even though we calculated the entropy or variance in each connected component
differently from the standard maximum entropy method and Otsu's method in image
segmentation, the philosophy remains the same as in these two popular methods.

5.5 Graph-Cut Based Image Segmentation and Laplacian Matrices

For a graph $G = (V, E)$, a cut is a partition on V such that $V = A \cup B$ and $A \cap B = \emptyset$.
If G is a weighted graph, then $w(e) = w(u, v)$ is the weight on edge $e = (u, v)$. We
need to then find a partition for A and B so that

$$\min\{cut(A, B) = \Sigma_{u \in A, v \in B} w(u, v)\} \tag{5.10}$$

is a minimum. This is called a minimum cut problem. Comparing the graph-cut with image segmentation, we can see that the minimum cut will segment an image into two components.

However, there are so many data points in a picture and some neighborhoods will have exactly the same values if we use the weight graph for all points in the pictures. For instance, in $|V| = 1024 \times 1024 = 2^{20}$, the graph for E and w may be $|E| = |V|^2 = 2^{40}$ in size. This may require BigData or cloud computing techniques to process.

Thus, a technique of solving this issue begins with safely partitioning the picture into smaller regions, specifically near convex regions for building an easier new graph, where each node of the graph represents a similar homogeneous region that is treated as one big pixel. Therefore, the minimum cut for big pixels will be much faster.

Shi and Malik invented a method called the normalized cut. It almost always finds the part that contains fewer vertices. Normalized cut will balance the size factor of the cut [34]. We present a brief introduction to this technology, which is very popular today. The normalized cut defines a measure as follows:

$$N\_cut(A, B) = \frac{cut(A, B)}{Assoc(A, V)} + \frac{cut(B, A)}{Assoc(B, V)}, \tag{5.11}$$

where $Assoc(S, V) = \Sigma_{u \in S, v \in V} w(u, v)$ is the total "association" between S and V. $cut(A, B)$ and $cut(B, A)$ may differ if we consider direct graphs. It is proven that finding the smallest $N - cut(A, B)$ is an NP-hard problem. We can only find its approximate solution in a reasonable time. This method involves an excellent graph theory tool called the Laplacian matrix along with the eigenvector to obtain this approximation of the solution.

Let D be a diagonal matrix with $d_{ii} = \Sigma_j w_{ij}$ and $d_{ij} = 0$ if $i \neq j$ and $W = [w_{ij}]$. The Laplacian matrix is $(D - W)$.

Shi and Malik showed that X is the binary membership function of a set S with a value of 1 for the component-index number in S and a value of -1 if the component-index number is not in S [34]. If $cut(S, V - S)$ is the normalized cut, then

$$\frac{X^T (D - W) X}{X^T D X}$$

is minimized with the condition $X^T D\mathbf{1} = 0$. Therefore, solving the following eigenvalue equation will provide a solution for this problem.

$$(D - W)X = \mu D X$$

The second smallest μ value (the smallest is zero) indicates that its eigenvector X is the smallest segment when X is translated into a binary vector.

If D is a diagonal matrix with $d_{ii} = \Sigma j w_{ij}$, then $d_{ij} = 0$ if $i \neq j$ and $W = [w_{ij}]$.

The normalized cut method is an very elegant method in terms of applying graph theory and linear algebra to image segmentation. However, it is still quite slow in performance. Ren and Malik improved this algorithm using a method similar to the method of finding the similarity of big pixels [31]. Scientists at Canon Incorporated observed that it is related to the λ-connected method by Chen [4, 41]. In 3D or other massive searches, when the data points cannot fit into a single computer, the BigData technology can be used with big pixels or big surface cells.

When G is not a weighted graph, $W = [w_{ij}]$ is the adjacency matrix. In such a case, $(D - W)$ is a standard form of the Laplacian matrix. The Laplacian matrix has many applications in data processing and numerical analysis. For instance, to use the difference method to solve a partial differential equation, we will need to use the Laplacian matrix [1, 34]. It can also be used as the local linear embedding technique in manifold learning [33]. The Laplacian matrix is positive semidefinite so its eigenvalues are always greater than or equal to 0.

5.6 Segmentation, BigData, and Subspace Clustering

Future image processing including segmentation must involve cloud computing. For instance, a large volume 3D image may need to be put into multiple machines for processing. A video clip can also be viewed as a 3D image where one z-axis is the time recording. Object tracking in this case is the online 3D image segmentation and recognition problem. Our current technology can mainly track one or several simple objects. We can say that many 3D image processing problems are BigData problems.

In this section, we will first present a method that can segment an image faster than the number of pixels the image has when the image is stored in a quadtree format. Second, we will generalize the problem of subspace clustering where we do consider not only segmenting the linear space but also submanifold clustering.

5.6.1 Quadtree and Octree Based Image Segmentation

In this section, we introduce a fast algorithm for segmenting/classifying 2D images or 2D spatial data. The data is stored in quadtree or octree formats [11]. We will use topological and graph theory properties of spatial data and images to speed up the segmentation process.

A key idea of this method is to perform a segmentation process without restoring the data (image) frames. In other words, this segmentation will be done in a virtual or abstracted manner. Based on local connectedness and value-homogeneity, we will apply λ-connected segmentation and the mean-based region growing segmentation.

Our algorithms will make the segmentation process much faster by not decoding the quadtree indexing code before the segmentation.

The new algorithm for stream data will modify the boundaries of the segments in previous frames to predict the segments in upcoming frames. This could lead to the widespread use of segmentation technology for computer vision and geo-data processing, medical image processing, object tracking, geometrical simulation, database application, data-mining, and multi-dimensional data sets.

Quadtree and octree representations are commonly used in medical imaging and spatial databases to compress data [11, 29, 36]. We have introduced quadtree in Chap. 2.

The advantages to use quadtrees are: (1) Searching for a data object only requires $O(logn)$ time, (2) There is a very small number of squared blocks (cells, big-pixels) in general so we can develop a fast tracking algorithm for boundary extraction, and (3) Each leaf can be treated as a surface-cell so that, using a similar technique, we can get octree representation of the solid object.

A compressed image represented in the quadtree will have a leaf index and its value [9]. An image is split into four quadrants, namely Q_0, Q_1, Q_2, and Q_3, and they represent the upper-left quadrant, upper-right quadrant, lower-right quadrant, and lower-left quadrant, respectively. A specific format is used to describe the structure of the compressed image in quadtree representation. For example, $(Null, 0)$ means that the entire image is filled with "0", $(< 3 >, 128)$ means that the lower-left quadrant is filled with "128", and $(< 2 >< 1 >, 255)$ means that the upper-right quadrant and the lower-right quadrant of the image is filled with "255". In this example, the leaf size may be computed as: $\frac{n}{2^2}$ where n is the length of image.

Typical image segmentation must go through each point so the time complexity must be at least $O(n^2)$ [29], where n is the length of the image and we can assume $n = 2^k$. In the quadtree technique, a leaf $(< 2 >< 1 >, 255)$ will represent $n/2^2 \times n/2^2$ pixels.

Assume that the number of quadtree leaves is N. Then, the segmentation algorithm can be described by first defining the adjacency graph $G_Q = (V_Q, E_Q)$ for the quadtree stored image where each leaf is a node in G_Q. If u and v are two adjacent leaves in V_Q, then $(u, v) \in E_Q$. In a 2D image, there two neighborhood systems: the four-neighbor system where each point has only four neighbors and the eight-neighbor system where each point has eight neighbors [9, 11].

The following theorem was proven in [9, 11]:

Lemma 5.1. G_Q *has at most* 3N *edges for a four-neighbor system. In addition,* G_Q *has at most* 4N *edges for an eight-neighbor system.*

This result shows that we will have an even faster segmentation algorithm than a linear time algorithm.

Our investigations show that the segmentation can be done by directly using quadtree indexing code (without decoding) from the spatial database [11].

Fig. 5.5 λ-connected segmentation (λ = 112/255) without decoding the quadtree images

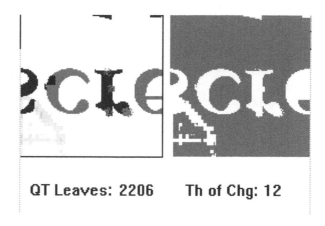

QT Leaves: 2206   Th of Chg: 12

Threshold segmentation is the most popular method. To make the discussion simple, we can assume that we are using a threshold or clip-level value to transform a gray-scale image into a binary image.

This method can naturally be used in the quadtree stored data set. We only need to examine the quadtree neighbors that are above the threshold (or between two thresholds in the multiple thresholding method.)

The results show we applied λ-connected segmentation without decoding the quadtree images in (Fig. 5.5).

The mean-based method is similar to the λ-connected method [16]. When we add a quadtree leaf to the segment, we want to check if the data is still within the range of the error limit (ϵ) with respect to the mean after the addition. However, it does not hold the transitivity property, meaning that the sequential order of adding leaves may affect the results. On the other hand, sometimes we need to split the leaf in order to satisfy the requirements. Because of this, we have to prioritize the order in which it is added. For example, the largest leaf can be added first in adjacency. Random adding can also be done if we do not care for a consistent result.

The problem with this method is that we usually do not test the pixels that are included, we only test the pixels that will be included. This reduces the segmentation time. However, it does not guarantee that the included pixels will still have the same properties as the individuals. As we discussed above, the sequence of the growing process can affect the results. For example, let us take an existing segment that tries to involve pixels/squares a and b. The order may change the results. This means the segment may reject a. However, if b merges first, then the new segment may accept a. This is why a split-and-merge segmentation method was proposed to set a certain order to have a relatively consistent result. As we can see, the statistical mean could not maintain the equivalence of segments [16].

To summarize, mean-based segmentation for quadtree data is not an equivalence relation partition. Different merging order may generate different results. The following order properties can be applied: (1) the largest region is applied first (2) the smallest difference (variation/standard derivation) is applied first. Real data results are shown below (Fig. 5.6):

Fig. 5.6 Mean-based
segmentation for a gray level
image without decoding the
quadtree images

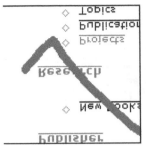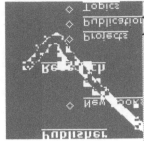

QT Leaves: 3256   Eps: 3

5.6.2 General Subspace Clustering

In mathematics, this problem is a related (algebraic) variety that is a union of some manifolds. A variety contains singular points. For example, two surfaces (or two two-manifolds) may intersect and have a curve as the intersection. The topology near this curve is not (homeomorphic to) a two-dimensional disk. Originally, variety was defined as the set of solutions of a system of polynomial equations.

In R^n, we like to identify all (sub-)manifolds. Storing these manifolds is much cheaper when n is very big. This problem is similar to subspace clustering defined in Chap. 1.

In pure mathematics, this problem is also related to Hodge's problem. Hodge's problem is also called the Hodge conjecture and is one of the seven Millennium Prize Problems (http://www.claymath.org/sites/default/files/hodge.pdf). This conjecture states: Let M be a non-singular complex projective manifold. Then every Hodge class on M is a linear combination with rational coefficients of the cohomology classes of complex subvarieties of M.

Roughly speaking, subspace clustering is to find a class of several systems of linear equations. The general subspace clustering is to find a class of several systems of polynomial equations. For example, if we could find a system of polynomial equations that is the best fit to the data set, then we can say that we can find the mathematical model of the data set.

Finding a connected component in discretized R^n is a more generalized procedure in image segmentation in high dimensions. This is a simple example of clustering. Assume that this component is k-manifold or k-complex, where $k < n$. We will have an object with reduced dimension.

The purpose is to find all submanifolds, where the union of these manifolds are all objects whose values are not zero. Each manifold is ideally a cluster in subspace clustering. Our task is to find a model that can describe such a cluster. Assume that we can find a system of polynomial equations that best fits one or more submanifolds (in terms of the union). Then, we have found the algebraic representation of the general subspace clustering. In other words, the Hodge problem could also have real-world examples in image segmentation (with modifications to transform R^n into a projective space).

5.7 Object Tracking in Video Images

Video is not usually saved as a sequence of image frames. It is often compressed. There is a difference between a real time video and a saved video. To store a sequence of image frames as a video clip is called encoding. For instance, we take a video clip from the cell phone, save it as a *.mov* file. This file uses MPeg4 as its compression technique. When we play this file, we need to decode each frame. In general, we need 30–120 frames per second to maintain good quality. Even though we only take 30 frames per second. We will have $24 \times 3600\,\mathrm{s} \times 30 = 2{,}592{,}000$ frames a day. Let us take a frame with $480 \times 640 = 307{,}200$ pixels and use three bytes per pixel as a colored image, then we will need 1 MB for each frame. Therefore, a camcorder will generate 2000 GB of data a day without compression. This is already near massive data storage scales. Walmart has over 5400 stores. The video data alone would already reach the 10 PB level if one store has only one camera at a doorway.

5.7.1 Videos, Compression, and Video Storing Formats

Video compression is a must when we deal with the sequence of image frames [16, 20, 30]. This is because a frame may contain the majority of the information in the next frame. For instance, in the evening, not so many people go to the store. Therefore, two consecutive frames are almost identical. A basic technique in image compression is to subtract the previous frame from the new frame. We only code the results of the subtraction called the residual. Since many pixels are zero in the residual, we can save a lot of space. The industry has developed several standards for video and image compression, including HVCC for instance.

Two common technologies for compression are called the cosine transformation, which uses half of the Fourier transform [See formula (5.1)], and the wavelet transformation. Usually, we use a one dimensional curve to code two dimensional information. See [16] for more information.

Here is the easiest formula for wavelet transformation: The wavelet transform is similar to the Fourier transform in terms of decomposing the signal. The wavelet transform has two advantages: (1) The wavelet transform can use many types of basic functions (basis) called wavelets. (2) Wavelet transform can perform simultaneous localization in both time and frequency domains [10, 22, 30]. A function $\psi(x)$ is called a wavelet. $\{\phi_{sk} : s, k \in I\}$ is called a basis ψ, if for integer s (scale factor) and k (time factor)

$$\phi_{sk}(x) = 2^{\frac{s}{2}} \psi(2^s x - k).$$

Let $\phi_{sk}(x)$ be basis functions where s is the scale and k is time. Let $f(x)$ be a function.

The integral wavelet transform is the integral transform defined as

$$W_\phi(\alpha, \beta) = \frac{1}{\sqrt{|\alpha|}} \int_{-\infty}^{\infty} \psi(\frac{x - \beta}{\beta}) f(x) dx \qquad (5.12)$$

where $\overline{g(x)}$ is the complex conjugate of $g(x)$. (For the complex number $z = a + ib$, its conjugate is $\bar{z} = a - ib$.)

We can represent a function $f(x)$ as

$$\mathbf{f}(x) = \sum_{s,k=-\infty}^{\infty} c_{sk} \phi_{sk}(x)$$

where

$$c_{sk} = W_\phi(2^{-s}, k2^{-s}).$$

The scale factor always decreases by 2. This method can only be used for multi-resolution or compression. People can select any indexes for the special purposes of compression. The simplest form of the wavelet is called the Haar wavelet.

An example of using Haar wavelets is as follows. For instance, in Haar wavelets, $H(t)$ can be described as

$$H(t) = \begin{cases} 1 & 0 \le t < 1/2, \\ -1 & 1/2 \le t < 1, \\ 0 & \text{otherwise.} \end{cases}$$

We have used Haar wavelets in meteorological data analysis for outlier tracking [22].

5.7.2 Object Tracking

In this section, we focus on the video tracking [21, 26, 37]. There are many video tracking methods, but we only introduce two of them here. The common technique now is to locate an object in a rectangular box, then the algorithm will trace the box that contains most of the component or object in the previous box in a frame [42].

Finding an object and tracking it can be different for compressed videos than for uncompressed videos with a compression, there is already some existing informa-

Fig. 5.7 Object tracking in videos [32]

tion about moving vectors (for some small image blocks, called macroblocks) that can be used. When the difference of two adjacent frames is too big, we will use the so called key frames directly in the video without compression. The following image frames will be compressed based on the key frame. However, for automobile driving software, the speed of finding an object is essential. There might not be enough time to do a compression.

As we can see, the segmentation of images is still the center of our concerns in video object tracking. This is because we need to first find an object that is our target. Then, the segments (components) in an image or video frame will be used to compare the predefined object we are looking for. Otherwise, we will isolate several objects that are moving to the next video frame. Those objects will be used as the predefined ones for the next search as well as determining the size and location of the rectangular region (called the bounding box) that holds the objects. Some fast trackers just use the feature vector of a bounding box to track the object in the next frame. See an example that is presented in Fig. 5.7 [32].

Smeulders et al presented a survey on the tracking methods for videos in [35]. A complete survey of research related to this topic up to 2006 can be found in [24]. Although there is a significant amount of research papers, compared to image segmentation, video tracking is still a relatively young research area.

5.7.3 Tracking Algorithms

The algorithm used to perform video tracking is called a tracker, which is used to match the representation of a target model in the current frame with its representation in the previous frame(s). Here, we use the representation of a target, meaning that we may use a feature vector to represent a component or a target. There are some noted trackers such as:

(1) Mean Shift Tracking [13]: This tracker makes match on histograms. It is very simple and elegant (comparing to use spatial information about the pixels or find shape-changes of the target).

(2) Normalized Cross-Correlation [2]: This tracker uses normalized cross-correlation to do direct target matching. It uses the intensity values in the initial bounding box as template. For each following frame, the tracker searches around the previous target position to select candidate windows (box). Compare each candidate window with the target template based on normalized cross-correlation. The best matching will be selected.

(3) Kalman Filter Tracking: The Kalman filter was originally invented to track the trajectory of moving objects in space. It is a two-stage filter running recursively based on an optimal Bayesian filter for Gaussian noise. The two stages are the predict step and the update step.

We now focus the method of mean-shift and its tracking method in videos. In the mean shift method, tracking a dense set of point data is based on the density function. To find the center point of each small window, we always move the current mean vector in the same direction as the maximum increase in density. This method is called the mean shift method.

The procedure guarantees the convergence of the point towards the points with the highest density in a designed window. Note that the gradient of the density function is zero at this point since the derivative of the highest point is zero. This process can start at a randomly selected point in the search space, this method can be used to find the backbone or framework of the data set. Mean shift can also be used to find the thinning (skeleton) of an object statistically.

Mathematically, mean shift can be used to find the centric trice of a cloud point data. We first need to assume a statistical model of density function, such as the Gaussian kernel: $Kernel(x - x_0) = e^{-c(x-x_0)^2}$ where x_0 is an initial estimated center where c is a predefined constant. Then the next mean will be determined by

$$m(x_0) = \frac{\sum_{x \in N(x_0)} Kernel(x - x_0)x}{\sum_{x \in N(x_0)} Kernel(x - x_0)} \tag{5.13}$$

where $N(x_0)$ is the neighborhood window of x_0, a set of points for which $Kernel(x_0) \neq 0$.

Then we will set the new center x_0 to be $m(x_0)$, and repeat the above calculation to get a trice of x_0 as the track of the mean shift.

The principle of the mean shift method is used by Comaniciu, Ramesh, and Meer [13, 14] for object tracking in videos. It is called the mean shift tracker, and is probably the most effective tracker.

Before we give the actual algorithm, we explain the principle of this elegant method below: In video tracking, the idea of the mean shift tracker is not going to trace the mass center of the distribution of the sample. This is because the data is everywhere in an image. For instance, we can describe an object as a pre-banding box. Everything (every pixel) is considered as part of the object. The key of this algorithm is to use the density distribution function or probability density function to represent this object. That can be a histogram of the object in gray-scale or color. This histogram will be the model for the next location prediction.

Let us think about the same binding box B (or a circular region) with center location y in both the previous frame Q and the present frame P, which are two consecutive frames. The histogram of Q is the model of the object. We can calculate the histogram of P with respect to B. Then, we want to calculate the "distance" between two histograms. The best way to find the distance between two histograms is related to the Bhattacharyya coefficient.

We use $h(Q)$ and $h(P)$ to represent the histograms at location y. $h(Q)$ can be viewed as a function or a vector. For instance, $Q = (q_0, \cdots, q_{255})$ will represent the gray-scale histogram where q_i is the intensity value on i (or in the interval of $(i-1, i]$). When we deal with a color image, we can use a triple length vector to represent (*red, green, blue*). In general, q_i, $i = 1, \cdots, m$ is the model or pattern at the original location y. $h(P) = \hat{P}(y)$ is a candidate of prediction in the new frame. Therefore, we can get the new mean shift center z using the Eq. (5.13) in P. We then compare the similarities between $\hat{q} = \{q_i\}$ and $\hat{P}(y)$ along with the similarities between q_i and $\hat{P}(z)$. B will be moved to the center at z when we do this calculation relating to z. We will change the center to z if the similarity value of $\hat{q}$ and $\hat{P}(z)$ is bigger. In implementation, this change will be at the middle location between z and y.

Such a similarity value is calculated by the Bhattacharyya coefficient,

$$\rho(\hat{P}(y), \hat{q}) = \Sigma_{u=1}^{m} \sqrt{p_i(y) \cdot q_i}. \tag{5.14}$$

We can think about $p_i(y)$ as the value of the histogram P. This equals the cosine of vectors $(\sqrt{p_1}, \cdot, \sqrt{p_m})$ and $(\sqrt{q_1}, \cdot, \sqrt{q_m})$. Therefore, maximizing this value ρ is maximizing the color match.

We can express the algorithm as the following. For more mathematical details, see [14, 15]. Note that $\hat{q}$ does not have to be a histogram. The algorithm was designed in [14]. The representation of the algorithm here was based on the lecture notes of Wee Kheng Leow at National University of Singapore and [14].

5.7.3.1 Machining Learning Algorithms

Another method often used is the machine learning algorithm. When we want to search for a specific person or specific plate number in a video clip, we can use neural networks to learn this object and then use the learned algorithm (a procedure coded) on each component or block in the video clip. Some other algorithms in machine learning can also be developed such as decision trees. Video tracking is still a relatively new research area, but some profound algorithms are expected to be developed in the future. An algorithm was presented in [32].

Algorithm 5.1 The mean shift tracking algorithm is as follows: Given $\{q_i\}, where i = 1, \cdots, m$, of a model, location y in the previous frame, and a small value ϵ as the moving length for each step, $\{q_i\}$ can be obtained by the histogram of colors in a box surrounding y in the previous frame. We want to find a location z, such that $\hat{P}(z)$ for $B(z)$ in the current frame has the maximum similarity to $\hat{P}(y)$ for $B(y)$ in the previous frame. (Note that box $B(y)$ does not necessarily appear, but it is good for our intuition. In fact, the calculation of $\{q_i\}, i = 1, \cdots, m$ already contains information on $B(y)$ or $B(z)$ for new center.)

Step 1.   Set the initial location to be y in the current frame to be the target. Compute $\hat{P}(y) = (p_1(y), \cdots, p_m(y))$ in the current frame, and compute $\rho(\hat{P}(y), \hat{q})$.

Step 2.   Assigning smaller weights w_j to pixels farther from the center, the calculation is based on the kernel in Equation (5.13). w_j will be used in the new center calculation.

Step 3.   Get the new location z using Equation (5.13).

Step 4.   Get $\hat{P}(z) = (p_1(z), \cdots, p_m(z))$ with new box $B(z)$ and $\rho(\hat{P}(z), \hat{q})$.

Step 5.   If $\rho(\hat{P}(z), \hat{q}) < \rho(\hat{P}(y), \hat{q})$ and $\|z - y\| > \epsilon$ (in the new frame), update z as $z \leftarrow \frac{1}{2}(y + z)$. Repeat Step 5.

Step 6.   Otherwise, when $y \leftarrow z$, assign the current frame to be the previous frame, and assign the next frame to be the current frame. Go to Step 2.

5.7.3.2 Online Object Tracking

Online object tracking is a new research area in video tracking [23, 35]. This problem can be specified to describe an object in the first frame of a video clip that we want to locate in subsequent frames with bounding boxes.

Online object tracking differs from typical video tracking in the following ways: (1) the variations of the object itself including incomplete data appearance, and (2) the complexity of the scene including the movement of cameras.

In [23], the authors developed an algorithm called tracking-learning-parsing (TLP) for online tracking. In the algorithm, the object is traced by a graph-based structure in a hierarchical and compositional model.

Even though the minimal spanning tree method is already very practical, techniques using segmentation based techniques are in development.

5.8 Future Concerns: BigData Related Image Segmentation

In general, images and videos are collections of raw data. Image analytics can help humans make decisions. Such analytics will add a great deal of value to the industry when a prescriptive resolution can be given, such as some of the methods mentioned in this chapter. Currently, researchers are concerned with emerging BigData images that contain advanced computational and scientific technologies in statistics, machine learning, operations research, and business logic or rules [39].

Mathematically, to segment an image is to partition it into several components. The partition is equal to an equivalence relation. The difference is that the segmentation allows some errors on the edge of each segment since almost every image contains noise.

It is definitely not possible to get a perfect image segmentation that is based on human interpretation. Different people will see or interpret an image differently.

The idea of rough sets, or sets that are λ-connected or normal connected, is good for finding upper and lower limits [5]. The question is how we find a reasonable control for these limits, a relation R or a λ value.

It is good that image segmentation can be treated in partition regions with different lambda-values for segmentation in different machines.

When we use Hadoop or MapReduce in image segmentation, we can pass different seeds to different locations of the array before passing them to different machines. The segmentation will be very fast in the region growing method.

References

1. M. Belkin, P. Niyogi, Laplacian eigenmaps for dimensionality reduction and data representation. Neural Comput. **15**, 1373–1396 (2003)
2. K. Briechle, U.D. Hanebeck, Template matching using fast normalized cross correlation, in *Proc. SPIE*, vol. 4387 (2001), pp. 95–102
3. T. Chan, L. Vese, Active contours without edges. IEEE Trans. Image Process. **10**(2), 266–277 (2001)
4. L. Chen, The λ-connected segmentation and the optimal algorithm for split-and-merge segmentation. Chin. J. Comput. **14**, 321–331 (1991)
5. L. Chen, λ-connected approximations for rough sets, in *Rough Sets and Current Trends in Computing*. Lecture Notes in Computer Science, vol 2457 (Springer, Heidelberg, 2002), pp. 572–577
6. L. Chen, *Discrete Surfaces and Manifolds: A Theory of Digital-Discrete Geometry and Topology* (S&P Computing, Rockville, 2004)
7. L. Chen, λ-measure for bone density connectivity, in *Proceedings of IEEE International Symposium on Industrial Electronics*, Montreal (2006), pp. 489–494
8. L. Chen, λ-connectedness determination for image segmentation, in *Proceedings of 2007, International Conference on Artificial Intelligence and Pattern Recognition* (2007), pp 71–79
9. L.M. Chen, Digital Functions and Data Reconstruction (Springer, New York, 2013)
10. L.M. Chen, Digital and Discrete Geometry: Theory and Algorithms, New York, Springer, 2014
11. L. Chen, H. Zhu and W. Cui, Very fast region-connected segmentation for spatial data: case study, in *IEEE conference on System, Man, and Cybernetics* (2006), pp 4001–4005
12. M. Cheriet, J.N. Said, C.Y. Suen, A recursive thresholding technique for image segmentation. IEEE Trans. Image Process. **7**(6), 918–921 (1998)
13. D. Comaniciu, V. Ramesh, P. Meer, Real-time tracking of non-rigid objects using mean shift, in *Proc. IEEE CVPR*, Hilton Head (2000)
14. D. Comaniciu, V. Ramesh, P. Meer, Kernel-based object tracking, IEEE Trans. Pattern Anal. Mach. Intell. **25**(5), 564–575 (2003)
15. T.H. Cormen, C.E. Leiserson, R.L. Rivest, *Introduction to Algorithms* (MIT Press, Cambridge, 1993)
16. R.C. Gonzalez, R. Wood, *Digital Image Processing* (Addison-Wesley, Reading, 1993)

17. L. Hertz, R.W. Schafer, Multilevel thresholding using edge matching. Comput. Vis. Graph. Image Process. **44**, 279–295 (1988)
18. S.L. Horowitz, T. Pavlidis, Picture segmentation by a tree traversal algorithm. J. Assoc. Comput. Mach. **23**, 368–388 (1976)
19. J.N. Kapur, P.K. Sahoo, A.K.C. Wong, A new method of gray level picture thresholding using the entropy of the histogram. Comput. Vis. Graph. Image Process. **29**, 273–285 (1985)
20. R. Klette, *A Concise Computer Vision* (Springer, London, 2012)
21. B. Li, R. Chellappa, Simultaneous tracking and verification via sequential posterior estimation, in *Proc. IEEE Conf. on Computer Vision and Pattern Recognition*, vol. II, Hilton Head (2000), pp. 110–117
22. C.-T. Lu, Y. Kou, J. Zhao, L. Chen, Detecting and tracking region outliers in meteorological data. Inf. Sci. **177**, 1609–1632 (2007)
23. Y. Lu, T. Wu, S.-C. Zhu, Online Object Tracking, Learning and Parsing with And-Or Graphs, CVPR 2014. (yanglv@ucla.edu, tfwu@stat.ucla.edu, sczhu@stat.ucla.edu)
24. T. Moeslund, A. Hilton, V. Kruger, A survey of advances in vision-based human motion capture and analysis. Comput. Vis. Image Underst. **104**, 90–126 (2006) [A complete survey for the papers up to 2006]
25. D. Mumford, J. Shah, Optimal approximation by piecewise smooth functions and associated variational problems. Commun. Pure Appl. Math. **42**, 577–685 (1989)
26. H.T. Nguyen, A.W.M. Smeulders, Fast occluded object tracking by a robust appearance filter. IEEE Trans. Pattern Anal. Mach. Intell. **26**(8), 1099–1104 (2004)
27. N. Otsu, A threshold selection method from grey-level histograms. IEEE Trans. Syst. Man Cybern. **SMC-8**, 62–66 (1978)
28. N. Pal, S. Pal, A review on image segmentation techniques. Pattern Recogn. **26**, 1277–1294 (1993)
29. T. Pavilidis, *Algorithms for Graphics and Image Processing* (Computer Science Press, Rockville, 1982)
30. J.-C. Pinoli, *Mathematical Foundations of Image Processing and Analysis*, vols. 1, 2 (Wiley, Hoboken, 2014)
31. X. Ren, J. Malik, Learning a classification model for segmentation, in *Proc. IEEE International Conference on Computer Vision*, pp. 10–17 (2003)
32. D.A. Ross, J. Lim, R.S. Lin, Incremental learning for robust visual tracking. Int. J. Comput. Vis. **77**, 125–141 (2008)
33. S. Roweis, L. Saul, Nonlinear dimensionality reduction by locally linear embedding. Science **290**(5500), 2323–2326 (2001)
34. J. Shi, J. Malik, Normalized cuts and image segmentation. IEEE Trans. Pattern Anal. Mach. 712 Intell. **22**(8), 888–905 (2000)
35. A.W.M. Smeulders, D.M. Chu, R. Cucchiara, S. Calderara, A. Dehghan, M. Shah, Visual tracking: an experimental survey. IEEE Trans. Pattern Anal. Mach. Intell. **36**(7) (2014)
36. M. Spann, R. Wilson, A quad-tree approach to image segmentation which combines statistical and spatial information. Pattern Recogn. **18**, 257–269 (1985)
37. A. Teran, Real-time multi-target tracking: a study on color-texture covariance matrices and descriptor/operator switching. PhD. Thesis, Univ. of Paris Sud Paris XI (2013)
38. L. Tsai, F.T. Berkey, Ionogram analysis using fuzzy segmentation and connectedness techniques. Radio Sci. **35**(2), 1173–1186 (2000)
39. F. Venter, A. Stein, Images and videos: really big data Analytics, pp. 1–11, November/December, 2012. (http://www.analytics-magazine.org/november-december-2011/694-images-a-videos-really-big-data)
40. T. White, Hadoop: the definitive guide, O'Reilly Media; 4 edition (2015)
41. M. Yamanaka, M. Matsugu, Information processing apparatus, control method for information processing apparatus and storage medium US 8792725 B2, Canon Kabushiki Kaisha. http://www.google.com.ar/patents/US8792725
42. C. Yang, R. Duraiswami, L. Davis, Fast multiple object tracking via a hierarchical particle filter, in *International Conference on Computer Vision*, Beijing, 15–21 October 2005

Chapter 6
Topological Data Analysis

Li M. Chen

Abstract Classical data processing uses pattern recognition methods such as classification for categorizing data. Such a method may involve a learning process. Modern data science also uses topological methods to find the structural features of data sets. In fact, topological methods should be the first step before the classification method is applied in most cases. Persistent homology is the most successful method for finding the topological structure of a discrete data set.

This chapter deals with topological data processing. We first introduce space triangulations and decompositions. Then, we discuss manifold learning and focus on persistent analysis. We give an overview of all topological methods but will focus on persistent data analysis.

6.1 Why Topology for Data Sets?

What is the difference between two data sets? This depends on what is the most important issue we care about these two data sets. When we talk about apples, the size or the number of apples is important. When we talk about automated car license plate recognition, we need to care about getting a good, clean image.

Netflix, an Internet based movie streaming and renting company, would want to predict revenues from a specific movie that targets a certain age group or demographic. We might ask: What is the volume (number of individuals) of people in an area? Can local internet servers in a city handle the movie stream? Is *city A* similar to *city B*?

We can classify the data into two categories for a certain movie: (1) people who are definitely interested and (2) people who are not interested.

We have discussed the NetFlix matrix completion problem in Chap. 2. Now the problem discussed here is more general especially in topological aspects. This is because that matrix completion is filling the holes using subspace, the completion of filling has so many options, a particular one might not reflect the actual data.

L.M. Chen (✉)
The University of the District of Columbia, Washington, DC, USA
e-mail: lchen@udc.edu

© Springer International Publishing Switzerland 2015
L.M. Chen et al., *Mathematical Problems in Data Science*,
DOI 10.1007/978-3-319-25127-1_6

On the other hand, holes in the matrix might be real. There are a group of people who will not watch a particular movie, we need to identify holes not to fill the holes.

Considering all the factors, Netflix will obtain a multidimensional cloud data set. Now, before Netflix purchases the rights to show a movie nationwide, they may test the movie in a small city. When the company has received the data for who or which families are watching the movie, the data will be translated into a set of vectors.

Even if similarities between cities exist, they may not be exactly the same. We can use deformation at the top of the training data to predict the topological structure of the large data set for a big city. Netflix can find the appropriate volume of people who are going to order the internet movie stream.

Another example is more practical and is used in medical imaging specifically bone data imaging. There are many holes in a particular part of a bone. The number of holes can indicate the strength of the bone. How do we calculate the correct number of holes [4, 5].

The third problem is related to wireless networking. We know we want to build several tower stations in a small city to handle the communication of cell phone users. Every stations will have a radio power that will cover certain area, usually circular area. We want to know if there is a hole where no station will reach it [6, 7]? These type of questions require us to find the structure or shape of the data. The topological structure is just the basic structure of these problems. We might be able to find valuable problems relating to topology in other business related data mining [30].

6.2 Concepts: Cloud Data, Decomposition, Simplex, Complex, and Topology

Cloud data, sometimes called scattered data, is a set of data points that are randomly arranged. They are usually dense, meaning that they have a lot of data points. Cloud computing, as well as cloud data computing, is highly related to networking.

A Problem of Wireless Networking Let us assume that we have a set of sites where each site has a tower station for network communication. When we drive a car from one location of a site to another, how do we handle the location change in terms of communication?

In fact, networking experts have already designed a way to switch a user (a cellphone) from one tower station to another by measuring distances. We always select the nearest host station to the user to handle communication.

The method in mathematics is called the Voronoi diagram. The method prepartitions the space into polygons. When a user enters in a specific polygon called Voronoi region, then the corresponding tower station takes over the service. This question is referred to in Chap. 1. Section 1.5.

The Voronoi Diagram The Voronoi diagram partitions a plane into polygons, where each polygon contains a sample point called a site; a point x is inside a specific polygon containing site p if x is closer to p than to any other site [4, 9, 35].

The Voronoi diagram method is particularly important to science and engineering. This partition is made based on the closest distance from the given site compared to other sites. The dual diagram of the Voronoi decomposition is a triangulation of the domain, which is called Delaunay triangulation.

Delaunay triangulation is the most popular form among different types of triangulation. See Fig. 3.4 from [4]. We will present the algorithm of Delaunay triangulation in the next section.

A space usually can be partitioned into smaller pieces. This is called a decomposition. Image segmentation is a type of decomposition. If these pieces are all triangles, we say this is a triangulation. We usually use polygons for more general decompositions. A triangle is called a 2D simplex (2-simplex). In a triangulation, two triangles share an edge (1-simplex), a point (0-simplex), or an empty set. A collection $\mathcal{K}$ of 2-simplexes, 1-simplexes, 0-simplexes (with the empty set) is called a simplicial complex if (a) the intersection of any two simplexes is also in $\mathcal{K}$, and (b) each edge or points of a simplex is an element in $\mathcal{K}$.

A topological space is a pair of $M = (X, \tau)$ where X is a set and τ is a collection of subsets of X. M satisfies (1) X and $\emptyset$ are in τ, (2) The union of any number of members of τ is in τ, and (3) The intersection of finite number of members of τ is in τ. We can see that a simplicial complex (usually contains finite number of simplexes) can be regarded as a topological space. We can see that a decomposition induces a simplicial complex.

In many other cases especially in digital images, we use squares to decompose a space. A square or other shape is called a cell, so we will have cell complexes. A triangulation of a space forms the simplicial complexes that is a foundation of Combinatorial Topology [4, 9, 13]. The formal definition of topology can be found in [4, 13].

6.3 Algorithmic Geometry and Topology

As we discussed in the last section, partitioning a 2D region into triangles is called triangulation. We can also partition a region into polygons where each polygon can be viewed as a cell. Therefore, we will have a cell complex. In higher dimensions, this is called a polyhedron. The polyhedron decompositions such as Voronoi diagrams are specific research areas in computational geometry also called algorithmic geometry. Cell complexes are in the research area of topology and have a long history in mathematics [13].

A complex that is usually a finite topology is a collection of cells in different dimensions. For the k-complex, M is the complex whose biggest dimension is k. If for each point (0-cell) in M, we have a group of k-cells containing this point that is homeomorphic to k-dimensional Euclidean space, then this complex is called a k-manifold.

Computational topology studies the problems related to manifolds in computing including algorithm design and practical applications [9, 34].

In this section, we discuss a classic problem of algorithmic geometry that is related to space decomposition and a relatively new problem that is called manifold learning.

6.3.1 Algorithms for Delaunay Triangulations and Voronoi Diagrams

Mathematically, the method of the most meaningful decompositions is the Voronoi Diagram: Given n points (called sites) on a plane, we partition a space into several regions. Any point in the region is closer to the site in its own region. This type of region is called the Voronoi region.

For a new point x, find the closest site to x. In other words, if we have a new point x and try to find the closest site to the new point, we only need to decide which Voronoi region contains x. This problem is called the nearest neighbor problem as we discussed in Chap. 2. That is also used in pattern recognition.

Fortune found an optimum algorithm for Voronoi diagrams with time complexity $O(n \log n)$, but it is difficult to implement [4]. Here, we first design an algorithm for Delaunay triangulation. This relatively simple algorithm is called Bowyer–Watson algorithm for the Delaunay triangulation .

Let P_i, $i = 1, \cdots, n$ be n sites. A Voronoi region is bounded by edges and vertices. The vertex of the region is called the Voronoi point. We know that: (1) An edge of the Voronoi region must have the property that each point on the edge must be equal distance to the two sites. (2) A Voronoi point must have the property of being equidistant to the three sites.

This means there must be a circle containing these three sites centered at each Voronoi point. This circle is the circumcircles of the triangle containing all three site points. Such a triangle is called a Delaunay triangle.

Another definition of Delaunay triangulation is as follows: no circumcircle of any triangle contains a site. The following Bowyer–Watson algorithm is designed based on this fact. The Bowyer–Watson algorithm is one of the most commonly used algorithms for this problem. This method can be used for computing the Delaunay triangulation of a finite set of points in any number of dimensions. After the Delaunay triangulation is completed, we can obtain a Voronoi diagram of these points by getting the dual graph of the Delaunay triangles. See Fig. 6.1 [4].

Algorithm 6.1. The Bowyer–Watson algorithm is incremental in that the algorithm works by adding one point at a time to a valid Delaunay triangulation of a subset of the desired points. Then, it works in the new subset, adding points to reconstruct the new Delaunay triangulation.

Step 1   Start with three points of the set. We make the first triangle by linking three points with three edges.

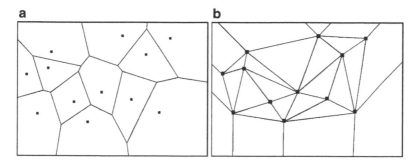

Fig. 6.1 Voronoi and Delaunay diagrams: (**a**) a Voronoi diagram; (**b**) Delaunay triangulation

Step 2   Insert a new point P. Draw the circumcircle of each existing triangle. If any of those circumcircles contains the new point P, the triangle will be marked as invalid.

Step 3   Remove all invalid triangles. This process will leave a convex polygon hole that contains the new point.

Step 4   Link the new point to each corner point of the convex polygon to form a new triangulation.

This algorithm is not the fastest one, and it runs in $O(n\sqrt{(n)})$ time. To get the Voronoi diagram from the Delaunay triangulation, the key is to link the centers of the circumcircles such that two corresponding triangles share an edge. For more details of this algorithm, see [4, 9, 35].

6.3.2 Manifold Learning on Cloud Data

Finding a surface from a 3D data volume is a very challenging job. However, if the data in the volume is continuous in terms of values, we can find an iso-surface, meaning that the value at each data point is the same. Then we can easily get the data set by searching for the same value in its neighborhood.

6.3.2.1 Real Problems Related to Manifold Learning

Isosurface Determination in High Dimensional Space In meteorology data processing, the researcher usually uses sensors to detect the temperatures or humidity data in the air. So reconstructing the shape from this type of sampling data points is not easy when we think the data is discrete points. In the past, we have some way that can draw a contour map. However, in such a case, we usually have a set of dense samples. We just need to extract the surface on which each point has the same or similar value called an isosurface. However if the data samples are shown

Fig. 6.2 Manifold learning
on cloud data [4, 17]

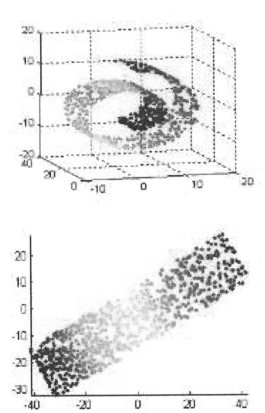

like Fig. 6.2a, it is hard to know the 2D neighborhood of a point in a desired surface
(e.g. an isosurface). Manifold learning is a way to find such a surface in a high
dimensional space.

Object Deformation Sequencing in a Set of Images Given a set of images, see
Fig. 6.3, can we determine a subset of images that are most likely in a deformation
sequence? In this problem, we will transfer the images or objects to feature vectors,
we want to find a smooth curve(s) in the space that holds these vectors (Fig. 6.4).
We then extract the curves and corresponding images. Such a smooth curve is a (1D)
manifold. This is also a good example in manifold learning.

Manifold learning here is to identify a lower dimensional manifold where most
of the data will be located.

Manifold learning is always related to dimensionality reduction. For instance, we
want to find a surface, a 2D object, from 3D space.

In Fig. 6.2, a 2D Swiss roll that was embedded in a 3D space [4, 17]. The question
is how do we extract the information?

For this particular case, we want to know if the data points represent a sphere or
a spiral shape [17, 32] (Manifold learning).

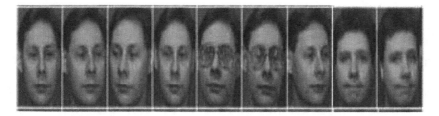

Fig. 6.3 Example of a set of facial images [32]

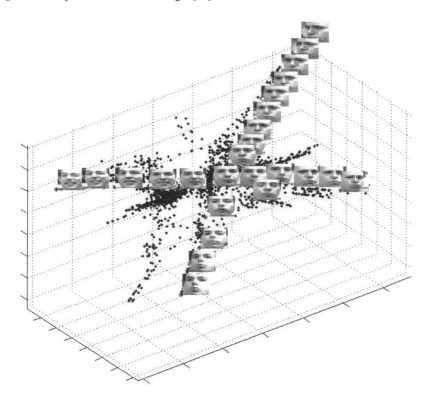

Fig. 6.4 Vector points of the set of facial images in 3D space [32]

Learning a data set that will most likely represent a random shape is a very difficult task. Therefore, there are many tasks to be studied in manifold learning [17, 32].

Two most popular methods for manifold learning are the Isomap and the kernel principal component analysis [4, 17]. The idea of the Isomap method is to use k-NN or MST to get the neighborhood information. Then, a graph is constructed with weight. After that, we use Dijkstra's algorithm to find the shortest path for each pair of points.

Let us present the Isomap algorithm below [4]:

Algorithm 6.2. The Isomap algorithm. Let S be a cloud data set.

Step 1   For each point $x \in S$, determine the neighbors of each point. We can use k-NN or MST to get the information.

Step 2   Construct the graph with the neighbors found in Step 1. The edge will be weighted by Euclidean distance.

Step 3   Calculate the shortest path between two vertices using Dijkstra's algorithm.

Step 4   Multidimensional scaling creates edges between two vertices. Cut off the data points beyond the clip level. Determine the local dimension of the data. This step is called lower-dimensional embedding.

Step 5   Compare the results from Step 4. Make a decision on the dimensions for all local neighborhoods. For instance, most local neighborhoods are 2D and so we make the 2D out put.

In Step 5, we can use principal component analysis to help us find the local dimensions [4, 17, 25]. In fast algorithm design, there may be some geometric data structures that can assist us to find new faster algorithms. The moving kernel will be determined by eigenvalues and eigenvectors. This method is referred to as the kernel principal component analysis.

In [32], a method called maximum variance unfolding (MVU) and the PCA method are used to find the image deforming sequence in Fig. 6.4. These results can be used to make better image reconstruction when we know the path of face change in deformation. See Fig. 6.5 [32]. More discussion can be found in [15]. This type of manifold learning has direct relationships to object tracking in videos. We have introduced it in Chap. 5. In manifold learning, we usually deal with very high dimensional data sets. For instance, we can treat a 32×32 picture as a $32 \times 32 = 1024$ dimensional vector. The calculation of this type is toward to a BigData type.

For massive images, cloud computing machines can be used to detect all possible directions of the pictures. This is also related to Bigdata analysis. Newly developed software, called SPARK, by researchers at Berkeley may play a significant role in this type of calculation [?]. In Chap. 8, Su et al. will discuss a special technique to manifold learning for data science.

Fig. 6.5 An example of ideal image reconstruction after finding the deformation line in Fig. 6.4

6.4 Persistent Homology and Data Analysis

As we discussed in the beginning of this chapter, individual sample points do not have a topological structure other than the discretely located points in space. However, human's interpretation makes a set of discrete points to have some meanings. The "best" interpretation is the structure we are looking for. A technology called persistent homology analysis was proposed to solve this problem [2, 8, 12]. It can be used to find the topological structure of a data set.

Let us have a set of cloud data, each point is an independent component. We can interpret each data point as a sample of a point, a small disk, or a cubical volume. We can also use this point represents a relatively bigger area, a bigger disk. When the size of the disks increases, the shape of the data sets may change its topological properties. For example, the data points may connect to be a object (component). See Fig. 6.6. The intuitive meaning of persistence is the topological property such as the number of components or the number of holes does not change in a period of time as the radius of disks changes.

While changing the value of radius, the homology or the number of holes does not always change, so the fit with the most unchanged holes is referred to as the best fit. This method is called persistent homology analysis [13, 36]. This method is becoming one of the major developments in topological data analysis.

6.4.1 Euler Characteristics and Homology Groups

In topology, homology usually indicates the number of holes in each dimension. We start at the Euler characteristic that is an invariant to a topological manifold. For any finite simplicial or cell complex, the Euler characteristic can be defined as the alternating sum

$$\chi = k_0 - k_1 + k_2 - k_3 + k_4 - k_5 \cdots, \tag{6.1}$$

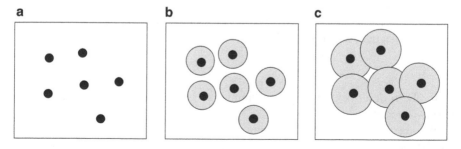

Fig. 6.6 Example of persistent analysis: (**a**) original data points, (**b**) the covered area by small disks centered at original points, and (**c**) the covered area by larger disks centered at original points; the topology is changed

where k_i denotes the number of i-cells in the complex. The detailed explanation of homology groups can be found in [13] and a concise introduction can be found in [4]. We only give a definition of homology groups here.

Let C_i be a group generated by i-cells in M. It is called a chain group where each element is in the form of $\Sigma_{A_i} n_i \cdot A_i$, where A_i is an i-cell and n_i is an integer. C_i is an Abelian group.

Let ∂_i be the boundary operator from C_i to C_{i-1}. The boundary operator only sends the i-cell to its boundary (with the direction already defined). We can see that the boundary of an i-cell is an "$(i - 1)$-cycle." The combination of an i-cell, for instance $kA + k'B$, will map to an element in C_{i-1} by ∂_i.

The ith homology group, H_i, is defined as

$$H_i(X) := Kernal(\partial_i)/Image(\partial_{i+1}), \tag{6.2}$$

For example, let a square $A = (a, b, c, d)$ be a 2-cell. Then, we have 4 edges, $(a, b), (b, c), (c, d), (d, a)$. So $\partial_2(A) = (+1)(a, b) + (+1)(b, c) + (+1)(c, d) + (+1)(d, a) = (+1)(a, b) + (+1)(b, c) + (+1)(c, d) + (-1)(a, d)$. $\partial_2(A)$ is an element in C_1. $Kernal(\partial_i)$ means the set of elements in C_i that maps to 0, the identity element in the Abelian group. In fact, all cycles in C_i will map to 0 in C_{i-1}. (The boundary of a cycle is empty, which means 0). $Image(\partial_{i+1})$ is the image of mapping ∂_{i+1}. They are boundaries of elements in C_{i+1}. Every boundary (in C_i) is a "cycle." Therefore, $Image(\partial_{i+1})$ is a subgroup of $Kernal(\partial_i)$.

To further explain, C_i is a group that contains all combinations of i-cells (with connected-sum as we can intuitively say). $Kernel(\partial_i)$ sends all cycles to $0 \in C_{i-1}$. $Image(\partial_{i+1})$ is a set of cycles in C_i that has filled $(i + 1)$-cells in K_{i+1} (the $(i + 1)$-sectionskeleton of M). This means that for the boundary of grouped $i + 1$-cells, the boundaries are in C_i. We want to pull them out of $Kernel(\partial_i)$ and only leave empty cycles in C_i. The removal means $Kernal(\partial_i)/Image(\partial_{i+1})$ boundaries.

The ith Betti number b_i is the rank of H_i. For example, if Z^n is a free Abelian group, then $\text{rank}(Z^n) = n$. We also have

$$\chi = b_0 - b_1 + b_2 - b_3 + \cdots, . \tag{6.3}$$

The Betti numbers are topological invariants that indicate the connectivity of simplicial or cell complexes. Let x be n-dimensional manifold. If X is closed and oriented, then the Poincare duality holds:

$$b_k = b_{n-k}.$$

In 2D, b_1 describes the maximum number of 1-cuts needed to separate a surface into two pieces.

We also have $\chi = 2 - 2g$, where g is the genus. Intuitively, b_i refers to the number of i-dimensional holes on a topological surface. So (1) b_0 is the number of connected components, (2) b_1 is the number of different type of one-dimensional holes (of a cycle in terms of deformation or homeomorphism), and (3) b_2 is the number of two-dimensional tunnels (cavities).

For a sphere S^n, where $n > 1$, we have $b_0 = b_n = 1$ and all other $b_i = 0$. For example, a torus has one connected component ($b_0 = 1$), two circular holes (b_1, one in the center and the other in the middle of the "donut"), and one two-dimensional cavity or tunnel (b_2, the inside of the "donut"), which yields Betti numbers. There is no 3D-cell so $b_i = 0$ for all $i \geq 3$.

Understanding homology groups is not a simple task. After extensive consideration and practice, we suggest methods for interpreting homology groups using the following rules:

1. Let M be an n-manifold or cell complex in general where b_0 is always the number of connected components. Now M can only be connected.
2. Draw an i-cycle σ_i (a manifold that is homomorphic to an i-sphere, S^i) in M, where $i < n$. If every other i-cycle can be deformed to the original σ_i, then $b_i = 0$ (if H_i is the remaining element in the group, then the group no longer exists).
3. If M is a closed n-manifold without a boundary, then $b_n = 1$. The boundary is the only "cycle" or closed boundary in n-dimension. $H_n = Kernel(\partial_n)/\{e = 0\}$, $\{e = 0\}$ is the identity group. M will map to 0 in C_{n-1} and is the generator of the group. For the same reason, if M contains k closed n-manifolds (where each manifold is "minimal"), then $b_n = k$. For instance, if M is 2-sphere, S^2, then $b_0 = 1$ and $b_1 = 0$ since every cycle on S^2 is deformable to another cycle (called homotopic) and $b_2 = 1$.
4. As we have said in (2), i-cycle σ_i and its deformed i-cycle will be treated the same. If there is another cycle and we cannot deform σ_i to it, then $b_i > 1$. If the example is for a torus, then we have $b_1 = 2$, but $b_0 = 1$ and $b_2 = 1$.
5. If M is a disk (or 2-ball), then $b_2 = 0$ since there is no 2-cycle. There are 1-cycles, but they are all homotopic (deformable to each other), so $b_1 = 1$. In general, for n-ball B^n, $b_0 = 1$ $b_1 = 0, \cdots, b_n = 0$.

The two following examples are very practical in application.

Example 6.1. For a graph $G = (V, E)$, if it is connected then $b_0 = 1$ and b_1 is the number of "minimal" cycles, which are called generators of H_1. A cycle is not "minimal" if it can be constructed by two (or more) "minimal" cycles using connected-sum.

Example 6.2. Let F be a connected image or picture in 2D. According to Rule (1), $b_0 = 1$. Based on Rule (5), $b_2 = 0$. The boundary of a hole is a cycle, which is minimal. According to Rule (3), b_1 is just the number of holes. We can also verify this using formula (6.1) and (6.3). Since $b_0 - b_1 + b_2 = k_0 - k_1 + k_2$, we have $b_1 = 1 + k_1 - k_0 - k_2$. For example, in Hather's book, for X_2 we have $k_0 = 2, k_1 = 4$, and $k_2 = 1$. There are two holes since $b_1 = 1 + k_1 - k_0 - k_2 = 1 + 4 - 2 - 1 = 2$. We can conclude that there are exactly two holes in the Euclidean 2D plane as it is shown in [13].

When we deal with applications in the following sections, we discuss how we can calculate and use algorithms that cannot be solved with the human eye.

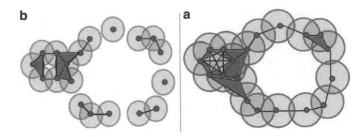

Fig. 6.7 Examples of Vietoris–Rips complex: (**a**), build VR complex using small disk, and (**b**) build VR complex using a relatively larger disk. [20]

6.4.2 Data Analysis Using Topology

Persistent data analysis has become more important in recent years. Many researchers have done some analysis on actual point cloud data sets. Algorithmically, it is not very efficient if we use small disks to cover the area and then do a homology calculation.

There is a method that is more practical called the Vietoris–Rips complex. It is related to the Cech complex that is made by the topological cover of $B(x, r)$.

Definition 6.1 (Vietoris–Rips Complex). Let X be a subset of a metric space with metric d. Choose a small real number $\epsilon > 0$. Construct a simplicial complex in the following way inductively:

(1) For each point in X, make it as a 0-simplex.
(2) For each pair $x_1, x_2 \in X$, make a 1-simplex ($[x_1, x_2]$) if $d(x_1, x_2) \le \epsilon$.
(3) For $x_1, x_2, \cdots, x_n \in X$, make an $(n - 1)$-simplex with vertices $x_1, x_2, \cdots, x_n$. Then, $d(xi, xj) \le \epsilon$ for all $0 \le i, j \le n$; that is, if all the points are within a distance of ϵ from each other.

This simplicial complex can also be denoted as $VR(X, \epsilon)$. The problem of this simplex is that we need extensive calculations to construct the complex, especially when X is a large set. It requires $O(2^{\{|X|\}})$ time complexity to determine such a simplex if we directly implement a simple algorithm as suggested by the definition, i.e. we check all possible simplexes. See Fig. 6.7 [20]. The construction process is called the Vietoris–Rips filtration. There has been software developed to build the complex [19, 20].

An algorithm for obtaining all possible Vietoris–Rips complex for each $r = \epsilon$ is not economic if the data set is very big. In the next section, we will find a very fast algorithm for 2D and 3D problems (Fig. 6.7).

6.5 Digital Topology Methods and Fast Implementation

Using a digital method, we can easily solve the problem of homology groups in 2D and 3D [4, 5]. This method is called digital geometry topology [4].

Digital geometry and topology was developed to provide a solid foundation for image processing and computer graphics. For 2D images, a connected component usually means an object in the image. A digital image can be viewed as an array; for simplicity, it can be viewed as a binary array with values only in $\{0, 1\}$. The value 0 indicates the background. Some theoretical development of digital topology related to digital surfaces are still under investigations [4, 16].

The topological structure of a connected component in 2D is essentially to find how many holes are in the component.

Thinking about the Vietoris–Rips or Cech complexes, these methods usually use $B(X, r)$ as the (open) covering set. If we use digital point plate (or square) $DB(X, r)$ to cover each point in X, then we will have the same topology. As we discussed above, for a digital component, we have $b_0 = 1$, $b_2 = 1$, and $b_1 = h$ where h is the number of holes in the component. If $DB(X, r)$ is a good cover, meaning that each intersection is simply connected (contractible), then the space made by $DB(X, r)$ has the same homology (groups) as the original manifold or complex sampled as the point set X.

The set $DB(X, r)$ refers to the digital cell complex. See [4] for details. The advantages of using $DB(X, r)$ instead of the Vietoris–Rips complex or $B(X, r)$ is that we can use the properties of digital topology to get a very fast algorithm in 2D and 3D.

To get $DB(X, r)$ algorithmically in a fast way is also interesting. We can use 6, 18, or 26 adjacency to determine whether the cubic cells are in $DB(X, r)$. For each r, we can obtain the homology groups or Betti numbers to get the persistent homology. In the following two subsections, we will introduce the digital methods.

For higher dimensional complexes, we still need to use the Vietoris–Rips complex. We would also need further research to archive the simplicity of computing.

6.5.1 2D Digital Holes and Betti Numbers

There is a very simple formula to get the number of holes in 2D digital space. 2D digital hole counting begins with the following.

Proposition 6.1. *For a connected component that has at least one 2-cell in 2D, we will have $b_0 = 1$, $b_2 = 1$ (at least one 2-cell in an image), and $b_1 =$ number of holes.*

This is because: (a) $H_0 = Z$ when the image is connected, (b) $H_2 = 0$ since ∂_3 does not exist, and (c) according to the definition, $\chi = \Sigma(-1)^i b_i = \Sigma(-1)^i k_i$. So $1 - b_1 + 0 = k_0 - k_1 + k_2$. For simplicity, let us use examples to verify this formula. See Fig. 6.5. We have 32 vertices, 48 edges, and 16 faces in Fig. 6.5a, so $k_0 = 32$, $k_1 = 48$, and $k_2 = 16$. Thus, $b_1 = 1 - (k_0 - k_1 + k_2) = 1$ in Fig. 6.5a.

In Fig. 6.5b, $k_0 = 32 + 21$, $k_1 = 48 + 34$, and $k_2 = 16 + 12$. Therefore, $b_1 = 2$ in Fig. 6.5b.

(Then, $H_1 = Z \times \cdots \times Z = Z^{b_1}$.)

Following $\chi = \Sigma(-1)^i b_i = \Sigma(-1)^i k_i$, this can be calculated as $b_1 = h$, where h is the number of holes (Fig. 6.8).

Therefore, using the digital method, we can easily find homology groups in 2D.

The number of holes and hole counting is important to determine whether two images are similar or completely different.

Let us look at the digital case of this problem. In Fig. 6.5, we defined In_p as the total number of corner points, each of which directs to the inside of the object. Likewise, Out_p is the number of total corner points, each of which directs to the outside of the object.

The outside boundary curve always has 4 more outward points than inward points. However, each of the inside cycle has 4 more inward (corner) points than outward points. We assume that M has h holes. Then, we will have $h + 1$ cycles, one of which would be outside boundary cycle B. Therefore,

Theorem 6.1.

$$h = 1 + (In_p - Out_p)/4. \tag{6.4}$$

The formal and topological proof requires more sophisticated knowledge in topology, see [4]. To examine the correctness of the formula, we can still use the examples shown in Fig. 6.5. In Fig. 6.5a, we have $In_p = 4$ and $Out_p = 4$.

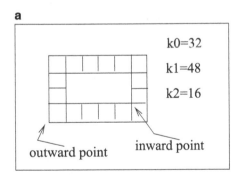
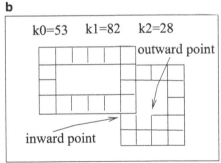

Fig. 6.8 Betti numbers and holes: (**a**), (**b**)

So $h = 1 + (In_p - Out_p)/4 = 1$. In Fig. 6.5b, $In_p = 11$ and $Out_p = 7$. So $h = 1 + (In_p - Out_p)/4 = 2$. We can see that formula (6.4) is much simpler than we just count k_0, k_1, and k_2.

In the image below, we can count the number of inward and outward points. We get $h = 1$.

6.5.2 3D Genus Computation

For details on 3D genus computation, see Chap. 14 in [4]. We will only present the formula and algorithm here [4, 5].

Theorem 6.2. *Let M be a closed 2D manifold in 3D space. The formula for genus is*

$$g = 1 + (|M_5| + 2 \cdot |M_6| - |M_3|)/8. \tag{6.5}$$

where M_i indicates the set of surface-points, each of which has i adjacent points on the surface.

This formula provides a type of topological invariants such as genus and homology groups for 3D image processing. We also design a linear time algorithm that determines such invariants for digital spaces in 3D.

Such computations have direct applications in medical imaging as they can be used to identify patterns in 3D imaging, especially for bone density calculation [4]. In [4], we discussed the implementation of the method and the applications of digital mean curvatures to 3D image classifications.

We now present a linear algorithm for finding homology groups in 3D. The actual real data calculation is shown in Figs. 6.9 and 6.10.

Algorithm 6.3. Let us assume that we have a connected set M that is a 3D digital manifold in 3D space.

Step 1.   Track the boundary of M, $S = \partial M$, which is a union of several closed surfaces. This algorithm only needs to visit all the points in M to see if the point is linked to a point outside of M. This point would be on the boundary.

Step 2.   Calculate the genus of each closed surface in ∂M using the method described in Sect. 2. We just need to count the number of neighbors on a surface. and put them in M_i using the formula (6.5) to obtain g.

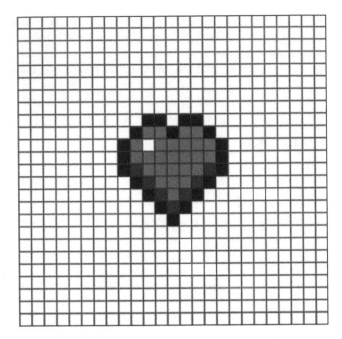

Fig. 6.9 Betti numbers and holes: $h = 1$ where $In_p = 18$ and $Out_p = 18$

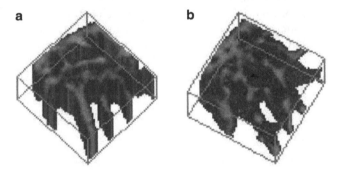

Fig. 6.10 3D genus calculation using a linear time algorithm in digital space: (**a**) g=6 and (**b**) g=10

Step 3.   we also can get H_0, H_1, H_2, and H_3. H_0 is Z. For H_1, we need to get $b_1(\partial M)$, which is the summation of the genus in all of the connected components in ∂M. H_2 is the number of components in ∂M and. H_3 is trivial.

6.6 New Developments in Persistent Homology and Analysis

Persistent homology for data analysis has been studied by many researchers in mathematics and computer science, Carlsson [2], Edelsbrunner and Harer [8, 9], Ghrist [6, 7, 12], and Zomorodian [34, 36], just to name a few.

Besides playing an important role in shape analysis, image processing, and computer vision, in recent years, persistent homology has been applied to many areas of data science including biology and medical science [26, 27] and sensor networks [6, 7]. Statistical informational analysis is also used in homology [10, 11]. Persistent homology has also been applied to natural language processing [33]. Introductory articles can be found in [2, 12, 31, 33].

Some theoretical analysis using theoretical learning theory and probabilistic learning are particularly interesting to computer theorists [22, 23, 31]. Software systems developed for homology groups can be found in [19, 29]. The principle of the algorithm design can be found in [19, 20].

6.6.1 Coverage of Sensor Networks Using Persistent Homology

Sensor Networks are the popular research area in electrical engineering, computer science, and information technology [1]. A fantastic method to solving a sensor networks coverage problem was developed using persistent homology [6, 7]. This problem was first considered by Grist, Problem 6.1 in this book.

In this problem, we want to know whether or not a set of sensors can cover a given area without leaving holes. However, we do not know the exact location of the sensors. We have a set of testing locations (users who have cell phones) that will be able to tell if the user is within a certain distance ($\leq r$) to a sensor i.

This problem was first considered by Silva and Grist in [6]. Tahbaz-Salehi and A Jadbabaie proposed a solving method that uses integer programming with persistent homology [28].

They have even developed a distributed algorithm that can localize coverage holes in a network of sensors without any metric information. This means we can use a computing cloud to accomplish the job. This is exactly the right algorithm for BigData networking. The implementation under Hadoop or SPARK architecture should not be very difficult.

This algorithm uses Vietoris–Rips complexes and finds the tightest (smallest or minimal) cycle encircling a hole. Since this problem is mostly in 2D, the question we can ask is the following:

Problem 6.1. Can we use the digital method for hole finding described in the above section to solve this problem in an even faster way? What is the time complexity of the new algorithm?

6.6.2 Statistics and Machine Learning Combined with Topological Analysis

A computational and statistical learning method was developed by Niyogi, Smale, and Weinberger for noise data where the homology was concerned [22, 23]. More interestingly, their method is related to a spectral learning algorithm based on a combinatorial Laplacian (introduced in Chap. 5) that is referred to as a sampling data-based simplicial complex [25].

This work is of particular interest in the areas of theoretical computer science and discrete mathematics. However, researchers usually emphasize algorithm analysis and not practical implementations.

Another aspect of statistical methods used in topological data analysis was in [11]. They focused on the practical uses of the statistical method in shape analysis.

However, when both sides realize the advantages of the other, new and more effective methods will be developed in the near future.

The implementation is always difficult for people in computer science and information technology. Since the mathematics behind homology requires an advanced mathematical background, the methods are not easily accessible to the broader public [33].

For instance, understanding Betti numbers is much more difficult than the term number of holes. However, the concept of holes is not precise to the homeomorphism of cycles. Training data scientists in algebraic topology is an issue that must be resolved in order for them to use such great technology.

To be frank, the concept of homology groups is a hard topic even for some mathematicians. At the same time, the material and concept is one of the toughest for a professor to teach in a popular way. For instance, how do we explain why a triangle without filling in 2D has $b_1 = 1$, but a triangle with filling has $b_1 = 0$? Intuitively, we can see that the triangle without filling has a hole, but the other does not. Using the theory of homology to derive the proof would be a difficult job.

6.7 Remarks: Topological Computing and Applications in the Future

In this section, we provide an author's view to some possible applications of topological data processing in the future. Finding the topological structure of a massive data set has gained much attention in mathematics and engineering. Space data and methodological data analysis require topological solutions.

A set of individual data points (cloud data) does not have a topological structure mathematically. However, since they are sampled from the real world, we can only get a discrete set. So the human interpretation of each unique cloud data set means something different. How do we find the best interpretation that matches the original object or event [12, 34]?

We know that each point is an individual event in sampling. However, we can interpret each data point as an area or volume (usually a disk or a ball), but we would not know how big the area or volume is. When the area is very small, the object may still remain not connected, but when the coverage area of a point is very big, the total area will be connected and become one connected object. So what would the radius be such that the filled area is the closest to the real world? This is also related to "data reconstruction" since a different radius chosen is the simple learning process.

6.7.1 λ-Connectedness and Topological Data Analysis

In the persistent homology method, we make a sample point grow in its volume with a radius r. When r changes from 0 to a big number, the data will change from individual data points to a large volume until fills the entire space.

The persistent homology method calculates the homology groups (number of holes in each dimension) for each r. It would make some sense that the same topology (homology groups) that covers most of r will be the primary topological structure of M, the data set.

The minimum spanning tree(MST) that finds the minimum value for r, to make all data samples connected. This we call D_c. The value of the largest distance of the pair of points in the space M will make the set to be simply connected, denoted by D_M. The smallest value that makes the set simply connected is called D_s, so we have $D_c \leq D_s \leq D_M$.

How do we find D_s using a fast search such as divide and conquer? Another way to represent the problem of persistent analysis in λ- connectivity is to build a reduction in the following way:

Get the geometric distance for all pairs, which we do not need to calculate beforehand. We put the largest value on the site i, then we put the value $f(j) = f(i) - d(i,j)$. We can see that every point will be adjacent to other points.

We could also use MST with normalization: Let us put point i as a point we are looking at. i will easily be λ-connected to its closer points. For any point j, j will be easily λ-connected to point k if j and k are in the same circle or sphere with respect to the center at site i. This is because the potential values on j and k are determined only by the distance to i even though j and k are very far from each other. The good asset about this setting is that if j and k are not λ-connected, then j and k are not close to each other.

As we stand at site i, we can do a partition using λ-connectedness. So we calculate homology based on the partitioned data. H^0 will be the number of the component.

We can also calculate the value based on j, and we can use it to separate the classes partitioned by i. This is a refinement process with regard to the existing one that is centered at i.

The advantage is that we can build λ-connectivity for each node and pass it to a cloud computer to do a focused calculation and combine the results.

Mathematically, the results will be the same and the complexity is not increased. We can do a clustering using kNN to only focus on the center points (to build a λ-connectedness for approximation).

Such a process will save a lot of time and make the calculations possible in cloud computing [14, 15, 30]. We only consider the original data points and do not need to work on the actual filling using disks. If a component contains a hole, we can use Minkowski's sum.

Based on point locations and radius, we can use triangulation, simplex, or digital methods for space coverage. Then, we can calculate the homology group. We first want a λ-connected partition because we want to treat the smaller subsets.

Another application of λ-connectedness is to use other factors, such as the shape when D_M, the largest distance between two points in the set, is small. We can also use a hierarchy for detailed topological analysis.

A λ-connected component indicates the topology. The homology of the lambda-connected segmentation is another kind of the bar-code of persistent analysis. For image segmentation, when the value of λ changes from 0 to 1, we can obtain the homology of the partition by considering each λ-connected components, $H(\lambda)$. What is the relationship between $H(\lambda)$ and the maximum spanning tree for λ we discussed in Chap. 3?

6.7.2   Hierarchy of λ-Connectedness and Topological Data Analysis

Data has shape and shape contains meaningful information. Topological data analysis (TDA) tries to find the topology of the data, whereas λ-connectedness can attach the geometric and statistical information to topological data analysis.

Each data cluster has its own shape and pattern, but what the relationship is among them is what we try to find when we make a classification of the classified data (or treat the cluster as an element for next level classification). λ-connectedness can measure the similarity of two objects. Let H_r is a persistent homology group for radius r. Let the set of H_{r_i} to be vertices of a graph, the λ-connectedness can be implemented among those groups that represent data partitions.

Basic topology of data has meaning, but the detailed analysis would provide more than topological information. More importantly, λ-connectedness uses the information obtained from TDA that is not totally independent from the TDA.

Using λ-connectedness in topological data analysis provides an extra parameter for assisting the existing topological method.

The value of the data recombines the average location. The shape factor or pattern factor is not enough in the next level of partitioning the geometric data information on the distance metrics.

We can use λ-connectedness to combine classifications from different data sets in different data forms.

For image processing, λ-connectedness can still apply in terms of partitioning into a Big Pixel.

This can even be applied to the deformed image where the deformation is a continuous move.

6.7.3 Topological Computing in Cloud Computers

Some companies do not want to offer all the data information to outside user, they can provide some necessary information for others to evaluate the company. This just like the outlines and statistics of a company. The "door" or "gate" for exchange information. But cannot provide the whole information. In image segmentation, we may need to only hock the boundary in the quadtree (split-and-merge), there is no need to know the entire quadtree.

When a company only wants to know certain information such as boundary information or doors not inside of a building, the topological structure can be calculated based on the boundary and inside data (secret inside calculation).

Just like in quadtree analysis, we only know the boundary of lines, and each station would have to take care its own calculations.

Data sharing has limitations in that data may not be detailed or a computer station cannot store the whole volume of data. Topological data analysis would need search and matching on boundaries.

Learning in λ-connectedness includes learning the local lambda value, quadtree analysis, and the kernel method or dynamically moving kernel methods.

λ-connectedness has advantages in that it can work with both a search method with topological characteristics and a classification method related to geometric data classification.

It is a general methodology for multitasking, independent to variability in BigData, and a fast implementation method for computer programming due to Depth First Search and Breadth First Search technology.

To get the shape of a component in the data sets, we must consider other features, such as average value and shape type. In summary, connectivity can be applied to find the topological structure and λ-connectivity can be used to find classification. This classification in geometry is the shape of the object.

Learning in λ-connectedness includes learning the local lambda value, quadtree analysis, and the kernel method or dynamically moving kernel methods.

Homology can find topological structure, but we still need to find the geometric structure of data points. For this purpose, λ-connectedness is a good candidate for data science.

Topological analysis is needed to model data sets that may not fill the entire space [14]. For instance, the rainfall during a hurricane will contain a centric area that does not have a lot of precipitation. Data reconstruction should consider the topology of the data sets.

When we consider a specific problem, what features are the most important? Average value, shape, center location, or other. λ-connectedness will provide a general tool for these problems.

For any data set, with or without initial value, when a basic partition or classification is made, we can build a function related to each class. Further classification or decomposition of the original data will use this information. For instance, for the computation of homology groups in terms of persistent analysis or Morse theory, we can use the connectivity of the data on the previous partition/classification. Not only do we need the simple distance metric but we also require additional information, though the distance metric (used in most persistent methods) is important.

For instance, in circle packing, for the deformation of the data, we want to find the deformation trice. This trice will represent the customer's marketing tendencies.

The λ-connectedness method will provide more refined and detailed techniques to the modern problem related to BigData. Again, for existing techniques, the λ-connectedness method will be highly applicable to reexamine a practical problem related to data science in finding and modeling a detailed structure of the problem. It could bring us one step closer to the satisfactory or desired solution of the problem and can be used to articulate the problem.

The following example provides an explanation to this problem: (1) A company owns a large forest. In regular years, there is a fertilizer plan with designed distribution. Each small circle will give a certain amount of fertilizer for a given cost. This area contains mountains and lakes. This year, due to strong winds from a valley near the forest and a long winter, the trees need more fertilizer.

Then, the planned picture image will be combined with sensor monitoring on the ground.

We want to use the ground sensors to do the first partition, and then we want to find the deformation curve of how much fertilizer is required. How do we deal with this problem?

A standard classification method can be used to find categories. The persistent analysis can be used to find the lakes that do not need to be treated. To find the deformation curve, the λ-connectedness method can help. The data sets would still need to be monitored during the winter.

Problem 6.2. How do we calculate homology groups based on connected sum #? We know that $\chi(M \times N) = \chi(M) \cdot \chi(N)$ and $\chi(M_1 \# M_2) = \chi(M_1) + \chi(M_2) - \chi(S^n)$. How do we use the similar method in homology groups calculation if we know the homology group of parts?

Problem 6.3. For a polyhedra or polytope in very large scales, how do we decompose it to be convex hulls? This problem was discussed between Dr. David Mount in UMD and the author.

Problem 6.4. Given a cloud set (can be very large), we will want to define a region (moveable) R, find Delaunay decomposition in R. Move R to cover all Space, R can be an N-ball. How do we find the local Delaunay decomposition with the merge considering intersections (smoothly)? The solution of this question will also solve the persistent homology problem. This problem was discussed between Dr. Feng Luo in Rutgers University and the author.

References

1. H.B. Cheng, Y.D. Yao, Power adaptation for multi-hop networks with end-to-end BER requirements. IEEE Trans. Veh. Technol. **59**(7), 3545–3554 (2010)
2. G. Carlsson, Topology and data. Bull. Am. Math. Soc. **46**(2), 255–308 (2009)
3. G. Carlsson, T. Ishkhanov, V. De Silva, A. Zomorodian, On the local behavior of spaces of natural images. Int. J. Comput. Vis. **76**(1), 1–12 (2008)
4. L.M. Chen, Digital and Discrete Geometry: Theory and Algorithms, New York, Springer, 2014.
5. L. Chen, Y. Rong, Digital topological method for computing genus and the Betti numbers. Topol. Appl. **157**(12), 1931–1936 (2010)
6. V. de Silva, R. Ghrist, Coverage in sensor networks via persistent homology. Algebr. Geom. Topol. **7**, 339–358 (2007)
7. V. de Silva, R. Ghrist, Homological sensor networks. Not. Am. Math. Soc. **54**, pp. 10–17 (2007)
8. H. Edelsbrunner, J.L. Harer, Persistent homology: a survey, in surveys on discrete and computational geometry, in *Twenty Years Later: AMS-IMS-SIAM Joint Summer Research Conference, June 18–22, 2006, Snowbird, Utah 453* (American Mathematical Society, Providence, RI, 2008), p. 257
9. H. Edelsbrunner, J. Harer, *Computational Topology: An Introduction*. Applied Mathematics (American Mathematical Society, Providence, RI, 2010)
10. B.T. Fasy et al., Confidence sets for persistence diagrams. Ann. Stat. **42**(6), 2301–2339 (2014)
11. J. Gamble, G. Heo, Exploring uses of persistent homology for statistical analysis of landmark-based shape data. J. Multivar. Anal. **101**(9), 2184–2199 (2010)
12. R. Ghrist, Barcodes: the persistent topology of data. Bull. Am. Math. Soc. **45**(1), 61 (2008)
13. A. Hatcher, *Algebraic Topology*, 1st edn. (Cambridge University Press, Cambridge, MA, 2001)
14. A. Holzinger, On topological data mining, in *Interactive Knowledge Discovery and Data Mining in Biomedical Informatics*. Lecture Notes in Computer Science, Springer, vol. 8401 (2014), pp. 331–356
15. E. Kokiopoulou, J. Chen, Y. Saad, Trace optimization and eigen-problems in dimension reduction methods. Numer. Linear Algebra Appl. **18**, 565–602 (2011)
16. T.Y. Kong, Minimal non-deletable sets and minimal non-codeletable sets in binary images. Theoret. Comput. Sci. **406**, 97–118 (2008)
17. T. Lin, H. Zha, Riemannian manifold learning. IEEE Trans. Pattern Anal. Mach. Intell. **30**(5), 796–809 (2008)
18. E. Munch, M. Shapiro, J. Harer, Failure filtrations for fenced sensor networks. Int. J. Robot. Res. **31**(9), 1044–1056 (2012)
19. V. Nanda, The perseus software project for rapid computation of persistent homology. http://www.sas.upenn.edu/~vnanda/perseus/index.html, 2012
20. V. Nanda, R. Sazdanovic, Simplicial models and topological inference in biological systems, in *Discrete and Topological Models Molecular Biology*, ed. by N. Jonoska, M. Saito (Springer, Berlin, 2014), pp. 109–141
21. M. Nicolau, A.J. Levine, G. Carlsson, Topology-based data analysis identifies a subgroup of breast cancers with a unique mutational profile and excellent survival. Proc. Natl. Acad. Sci. **108**(17), 7265–7270 (2011)
22. P. Niyogi, S. Smale, S. Weinberger, Finding the homology of submanifolds with high confidence from random samples. Discrete Comput. Geom. **39**, 419–441 (2008)

23. P. Niyogi, S. Smale, S. Weinberger, A topological view of unsupervised learning from noisy data. SIAM J. Comput. **20**, 646–663 (2011)
24. B. Rieck, H. Leitte, Persistent homology for the evaluation of dimensionality reduction schemes, in *Eurographics Conference on Visualization (EuroVis) 2015*, ed. by H. Carr, K. -L. Ma, G. Santucci, vol. 34(3) (2015)
25. A. Singer, From graph to manifold Laplacian: the convergence rate. Appl. Comput. Harmon. Anal. **21**, 128–134 (2006)
26. G. Singh, F. Mmoli, G. Carlsson, Topological methods for the analysis of high-dimensional data sets and 3-D object recognition, in *Eurographics Association Symposium on Point-Based Graphics 22 (The Eurographics Association, 2007)*
27. G. Singh, F. Memoli, T. Ishkhanov, G. Sapiro, G. Carlsson, D.L. Ringach, Topological analysis of population activity in visual cortex. J. Vis. **8**(8), 1–18 (2008)
28. A. Tahbaz-Salehi, A. Jadbabaie, Distributed coverage verification in sensor networks without location information. IEEE Trans. Autom. Control **55**(8), 1837–1849 (2010)
29. A. Tausz, M. Vejdemo-Johansson, H. Adams, Javaplex: A research software package for persistent (co)homology. Software available at http://code.google.com/javaplex. (2011)
30. W. van der Aalst, *Process Mining: Discovery, Conformance and Enhancement of Business Processes* (Springer, Berlin, 2011)
31. S. Weinberger, What is ... Persistent Homology? Not. AMS 36–39 (2011)
32. K.Q. Weinberger, L.K. Saul, An introduction to nonlinear dimensionality reduction by maximum variance unfolding, in *Proceedings of the Twenty First National Conference on Artificial Intelligence (AAAI-06)*, Cambridge, MA (2006), pp. 1683–1686
33. X. Zhu, Persistent homology: an introduction and a new text representation for natural language processing, in *The 23rd International Joint Conference on Artificial Intelligence (IJCAI)* (2013), pp. 1953–1959
34. A.J. Zomorodian, Computing and comprehending topology: persistence and hierarchical Morse complexes. Ph.D. thesis, University of Illinois at Urbana-Champaign, 2001
35. A. Zomorodian, *Topology for Computing* (Cambridge University Press, Cambridge, MA, 2005)
36. A. Zomorodian, G. Carlsson, Computing persistent homology. Discrete Comput. Geom. **33**(2), 249–274 (2005)

Chapter 7
Monte Carlo Methods and Their Applications in Big Data Analysis

Hao Ji and Yaohang Li

Abstract Due to recent emergence of many big data problems, Monte Carlo methods tend to become powerful tools for analyzing big data sets. In this chapter, we first review the fundamental principles of Monte Carlo methods. Then, we describe several popular variance reduction techniques, including stratified sampling, control variates, antithetic variates, and importance sampling, to improve Monte Carlo sampling efficiency. Finally, application examples of estimation of sum, Monte Carlo linear solver, image recovery, matrix multiplication, and low-rank approximation are shown as case studies to demonstrate the effectiveness of Monte Carlo methods in data analysis.

7.1 Introduction

Numerical methods known as Monte Carlo methods can be loosely defined in general terms to be any methods that rely on random sampling to estimate the solutions. Monte Carlo methods are often applied to problems which are either too complicated to be described by a mathematical model or whose parameter space is too large to be explored systematically.

Recent years have witnessed dramatic increase of data in many fields of science and engineering [1], due to the advancement of sensors, mobile devices, biotechnology, digital communication, and internet applications. These massive, continuously growing, complex, and diverse data sets are often referred to as the "big data." Analyzing big data demands cost-effective and innovative forms of information processing for enhanced insight and decision making. While many traditional, deterministic data analysis methods have great difficulties to scale to the massive big data sets, Monte Carlo methods, which are based on random sampling techniques, become important and powerful tools for big data applications.

H. Ji • Y. Li (✉)
Department of Computer Science, Old Dominion University, Norfolk, VA 23529, USA
e-mail: hji@cs.odu.edu; yaohang@cs.odu.edu

© Springer International Publishing Switzerland 2015   125
L.M. Chen et al., *Mathematical Problems in Data Science*,
DOI 10.1007/978-3-319-25127-1_7

Particularly, for certain big data sets beyond the computational capability of the most powerful supercomputers, Monte Carlo data analysis methods using random sampling are the only viable approaches.

This article starts with a basic description of the principles of Monte Carlo methods. It then discusses the variance reduction techniques, including stratified sampling, control variates, antithetic variates, and importance sampling, to design "smart" Monte Carlo. Examples of Monte Carlo methods on estimating sum of a large array, solving linear systems, multiplying matrix products, recovering missing matrix, and approximating low-rank matrices. Although in order to keep the presentation simple, these application examples are presented in a relatively small scale, the underlying Monte Carlo techniques can be effectively extended to big data sets.

7.2   The Basic of Monte Carlo

Monte Carlo methods provide approximate solutions to a variety of mathematical problems by random sampling. Let us take the numerical integration as an example, which constitutes a broad family of algorithms in numerical analysis. Suppose we want to calculate a one-dimensional definite numerical integral, $I = \int_a^b f(x)\,dx$. A common numerical integral method is to divide the one-dimensional interval into N subintervals and then to sum the area corresponding to each subinterval using either rectangular, trapezoidal, or Simpson's rules (Fig. 7.1a) [2]. Similarly, for two-dimensional intervals, the number of 2D subintervals becomes N^2 (Fig. 7.1b). In general, for d-dimensional integration problems, the d-dimensional space needs to be divided into N^d subintervals. For a not very high dimensional problem with $d = 20$ and $N = 100$, the total number of subintervals that need to be evaluated goes up to 10^{40}, which is unapproachable by most numerical integration algorithms. Mathematically, this is referred to as the "curse of dimensionality."

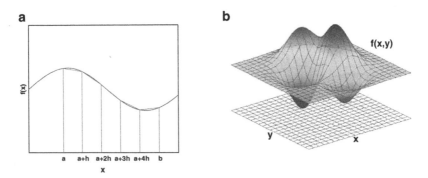

Fig. 7.1 Numerical integration using deterministic methods. (**a**) 1D integral. (**b**) 2D integral

In contrast, Monte Carlo methods estimate the integral by statistical sampling techniques [3]. Let us consider a one-dimensional integral $I_{0-1} = \int_0^1 f(x)\, dx$, which can be easily extended to a more general integral of $I = \int_a^b f(x)\, dx$. Suppose that the random variables $x_1, x_2, \cdots, x_N$ are drawn independently from the probability density function $p(x)$. A function F may be defined as

$$F = \sum_{i=1}^{N} f(x_i)\, p(x_i).$$

The expectation value of F becomes

$$E(F) = \int_0^1 f(x)\, p(x)\, dx.$$

The crude Monte Carlo integration method assumes that the probability density function $p(x)$ is uniform, i.e., the random samples $f(x_1), f(x_2), \cdots, f(x_N)$ are equally important, and then

$$E(F) = \int_a^b f(x)\, dx.$$

Correspondingly, the variance of F becomes

$$Var(F) = \frac{1}{N} \int_0^1 (f(x) - E(F))^2\, dx = \frac{1}{N}\sigma^2,$$

where σ^2 is the inherent variance of the integrant function $f(x)$. Clearly, we can find that the standard deviation of the estimator θ is $\sigma N^{-1/2}$. This means that as $N \to \infty$, the distribution of F narrows around its mean at the rate of $O\left(N^{-1/2}\right)$.

Now, let us extend the Monte Carlo integration method to a d-dimensional integral $I_d = \int_0^1 \cdots \int_0^1 f(x)\, dx$, the expectation of $F_d = \sum_{i=1}^{N} f(x_i)/N$ on uniformly distributed random variable vectors $x_1, x_2, \cdots, x_N$ becomes

$$E(F_d) = \int_0^1 \cdots \int_0^1 f(x)\, dx = I_d.$$

The variance of the estimator F_d is σ_d^2/N, where σ_d^2 is the inherent variance of the integrant function $f(x)$. If an integral function $f(x)$ is given, σ_d^2 is a constant and therefore, similar to one-dimensional integral, the convergence rate of Monte Carlo is $O\left(N^{-1/2}\right)$, which is independent of dimensionality.

In summary, compared to the deterministic numerical integration methods, whose convergence rate is $O(N^{-\alpha/d})$, where α is the algorithm related constant and d is the dimension, Monte Carlo integration method yields a convergence rate of $O(N^{-1/2})$ [4], which can somehow avoid the "curse of dimensionality."

Moreover, computations on each random samples are independent, which can be carried out in an embarrassingly parallel minor to harness the power of large-scale parallel and distributed computing architectures [5, 6]. On the other hand, the main disadvantage of Monte Carlo is that the convergence of Monte Carlo methods is very slow, roughly for every one digit of accuracy usually requiring 100 times more computations.

7.3 Variance Reduction

Crude Monte Carlo treats all random samples are equally important. In reality, we can often gain additional knowledge from the application domain, which can be taken advantage to come up with better estimators. Variance reduction is a procedure of deriving an alternative estimator to obtain a smaller variance than the crude Monte Carlo estimator and improve the precision of the Monte Carlo estimates for a given number of samples. In practical applications, a good estimator leading to million times more accurate than a bad one is not rarely seen. In this section, we describe some of the popular variance reduction techniques [3, 4], including stratified sampling, control variates, antithetic variates, and importance sampling. These variance reduction methods, if appropriately used, can significantly improve the efficiency of Monte Carlo methods in processing and analyzing big data sets.

7.3.1 Stratified Sampling

The inherent variance of the integral function $f(x)$ may vary significantly in different regions. The fundamental idea of stratified sampling is to apply more samples in the region with large variability and vice versa. Using the one-dimensional integral in interval $[0, 1]$ as an example, stratified sampling is to partition the interval $[0, 1]$ into several disjoint subinterval ranges, i.e., $[0, \alpha_1)$, $[\alpha_1, \alpha_2)$, $\cdots$, $[\alpha_{k-1}, \alpha_k)$, $[\alpha_k, 1]$. Then a Monte Carlo estimator $F_{\alpha_{i-1}, \alpha_i}$ is applied to each range $[\alpha_{i-1}, \alpha_i)$ separately so that

$$F_{\alpha_{i-1}, \alpha_i} = \sum_{j=1}^{n_i} (\alpha_i - \alpha_{i-1}) \frac{1}{n_i} f(x_j),$$

where x_j is a random sample in $[\alpha_{i-1}, \alpha_i)$ and n_i is the total number of samples. Clearly,

$$E\left(F_{\alpha_{i-1}, \alpha_i}\right) = \int_{\alpha_{i-1}}^{\alpha_i} f(x) dx.$$

The overall stratified sampling estimator $F_{stratified}$ is the combination of all estimators in different subinterval ranges

$$F_{stratified} = \sum_{i=1}^{k+1} F_{\alpha_{i-1},\alpha_i},$$

which is an unbiased estimator of the integral. The variance of the stratified sampling estimator becomes

$$Var\left(F_{stratified}\right) = \sum_{i=1}^{k+1} (\alpha_i - \alpha_{i-1}) \frac{1}{n_i} \sigma^2_{\alpha_{i-1},\alpha_i},$$

where $\sigma^2_{\alpha_{i-1},\alpha_i}$ is the variance of the integrant function $f(x)$ in subinterval $[\alpha_{i-1}, \alpha_i)$. This reveals the rationale of stratified sampling, which is to minimize $\frac{1}{n_i}\sigma^2_{\alpha_{i-1},\alpha_i}$ in each subinterval. If stratification is well carried out, the variance of stratified sampling will be less than that of crude Monte Carlo.

7.3.2 Control Variates

If there is another estimator which is positively correlated to the one to be evaluated and can be easily calculated with known expectation, it can be used as a control variate. For example, in the Monte Carlo integration example, the original integral $I_{0-1} = \int_0^1 f(x)dx$ can be rewritten into a summation of two parts

$$I_{0-1} = \int_0^1 g(x)dx + \int_0^1 (f(x) - g(x))\, dx,$$

where $g(x)$ is a simple function that can be either integrated theoretically or easily calculated and absorbs most of the variation of $f(x)$ by mimicking $f(x)$ as shown in Fig. 7.2, we only need to estimate the difference between $f(x) - g(x)$, which will result in variance reduction of the estimator.

7.3.3 Antithetic Variates

Contrast to the control variates method based on a strongly positively correlated estimator with F, the antithetic variates method takes advantage of an estimator F' that is strongly negatively correlated with F. Suppose that F is an unbiased estimator of a quantity, $E(F) = E(F')$, and $Cov(F, F') < 0$. Then, the antithetic estimator $F_{antithetic} = (F + F')/2$ is also an unbiased estimator of this quantity. The sampling variance of $F_{antithetic}$ becomes

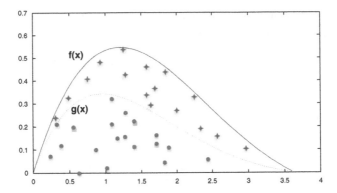

Fig. 7.2 Control variates approach. If $g(x)$ can be estimated theoretically, the sampling errors are limited to the area of $f(x) - g(x)$

$$Var\left(F_{antitheticc}\right) = \frac{Var\left(F\right)}{v} + \frac{Var\left(F'\right)}{4} + \frac{Cov\left(F, F'\right)}{2}.$$

Here $Cov(F, F') < 0$ due to negative correlation between F and F', which leads to overall smaller variance in $F_{antitheticc}$ than that of F and F'.

7.3.4 Importance Sampling

The importance sampling technology is often used in statistical resampling, which reduces variance by emphasizing the sampling on regions of interest. For example, by introducing a new proposal function $g(x)$, the original integral $I_{0-1} = \int_0^1 f(x)dx$ can be rewritten as

$$I_{0-1} = \int_0^1 \frac{f(x)}{g(x)} g(x)dx = \int_0^1 \frac{f(x)}{g(x)} dG(x),$$

where $G(x)$ is a cumulative density function (CDF). $f(x)/g(x)$ is called the likelihood ratio. With random samples drawn from a proposal distribution whose CDF is $G(x)$ instead of sampling from a uniform distribution, the variance of the importance sampling estimator $F_{importance\ sampling}$ becomes

$$Var\left(F_{importance\ sampling}\right) = \int_0^1 \left(\frac{f(x)}{g(x)} - E(F)\right)^2 dG(x).$$

A good likelihood ratio can result in a significant variance reduction.

In practice, assume that we know nothing about the target distribution at the very beginning, we may have to start from uniform sampling. However, after initial sampling, we have a better understanding of the target distribution, which results in a better proposal function. The resampling can be guided by the new proposal function and leads to a better approximation of the target distribution.

7.4 Examples of Monte Carlo Methods in Data Science

In this section, we present several application examples as case studies of using Monte Carlo methods for data analysis. Variance reduction techniques are applied to build "smart" Monte Carlo estimators to enhance sampling efficiency.

7.4.1 Case Study 1: Estimation of Sum

In many practical data analysis applications, we are often required to estimate the sum of a large data set. However, due to many reasons, we are not allowed to visit every element in the data set but want to get a good estimation. The following is an application example to estimate the overall salary expense in a company.

Consider a big, global company having 7140 employees falling in the following categories, 78 managers, 4020 engineers, 2008 salesmen, and 1034 technicians. Now we want to estimate the overall expense in employee salaries in this company. Due to cost as technical difficulty, we are only allowed to use 100 samples.

The simplest sampling method is the crude Monte Carlo using uniform sampling without considering different categories. The main problem of uniform sampling is ignoring the differences among categories, which will thus lead to a large estimation variance $460, 871K \pm 80, 712K$. More importantly, the percentage of managers is around 1 % in all employees. There is a high chance that the selected 100 samples may miss the manager category. Stratified sampling can better address this problem. We can calculate the number of samples falling into each category, specifically, $78/7140 \times 100 = 1$ sample in managers, $4020/7140 \times 100 = 56$ samples in engineers, $2008/7140 \times 100 = 28$ samples in salesmen, and $1034/7140 \times 100 = 15$ samples in technicians. As a result, stratified sampling yields better estimation $(500, 112K \pm 30, 147K)$ than that of uniform sampling.

If we have additional information, we can construct even better stratified sampling estimator. If we happen to know about the salary variation in each employee category (for example, from previous years' data), i.e., Managers ($200K \sim 500K$), Salesmen ($40K \sim 120K$), Engineers ($60K \sim 80K$), and Technicians ($50K \sim 70K$), we can take advantage of this information. We can find that the managers and salesmen have large variances, which deserves more samples while the engineers and technicians categories have small variances, where small number of samples will work. By reassigning the number of samples in different categories, 32 samples for Managers, 7 samples for Engineers, 59 samples for Salesmen, and 2 samples for Technicians, the overall estimated result becomes $520,066K \pm 10,113K$, which provides us more precise estimation than simple stratified sampling.

7.4.2 Case Study 2: Monte Carlo Linear Solver

Applying Monte Carlo sampling to estimate solutions in linear systems is originally proposed by Ulam and von Neumann and later described by Forsythe and Leibler in [7]. Considering a linear system of

$$x = Hx + b,$$

where H is an $n \times n$ non-singular matrix, b is the given constant vector, and x is the vector of unknowns. The fundamental idea of the Monte Carlo solver is to construct Markov chains by generating random walks to statistically sample the underlying Neumann series

$$I + H + H^2 + H^3 + \cdots$$

of the linear system [8]. The transition probabilities of the random walks are defined by a transition matrix P satisfying the following transition conditions:

$$P_{ij} \geq 0;$$

$$\sum_j P_{ij} \leq 1;$$

$$H_{ij} \neq 0 \rightarrow P_{ij} \neq 0,$$

and the termination probability T_i at row i is defined as

$$T_i = 1 - \sum_j P_{ij}.$$

Then, a random walk starting at i_0 and terminating after k steps is defined as

$$\gamma_k : r_0 \rightarrow r_1 \rightarrow r_2 \rightarrow \cdots \rightarrow r_k$$

where the integers $r_0, r_1, r_2, \cdots, r_k$ are the row indices of matrix H visited during the random walk.

As noted in existing literature [9], if the necessary and sufficient condition for the convergence of Monte Carlo solver holds, a random variable $X(\gamma_k)$ defined as

$$X(\gamma_k) = \frac{H_{r_0 r_1} H_{r_1 r_2} \cdots H_{r_{k-1} r_k}}{P_{r_0 r_1} P_{r_1 r_2} \cdots P_{r_{k-1} r_k}} b_{r_k} / T_{r_k}$$

is an unbiased estimator of component x_{r_0} in the unknown vector x.

Fig. 7.3 Comparison of Monte Carlo linear solver with different transition matrices

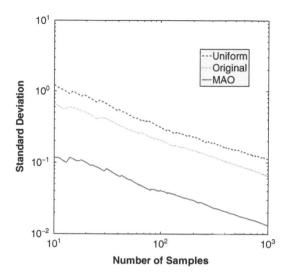

Variance reductions have been commonly employed in the Monte Carlo algorithms to improve the sampling efficiency. Let's consider a simple linear with

$$H = \begin{bmatrix} 0.1 & 0.45 & 0.225 \\ -0.15 & 0.1 & -0.3 \\ -0.18 & 0.36 & 0.1 \end{bmatrix} \text{ and } b = \begin{bmatrix} 0.225 \\ 1.35 \\ 0.72 \end{bmatrix}. \text{ Clearly, the exact solution is}$$

$$b = \begin{bmatrix} 1 \\ 1 \\ 1 \end{bmatrix}.$$

Figure 7.3 compares the convergence of Monte Carlo linear solvers using three sampling schemes listed in Table 7.1 with different transition matrices in terms of estimate value x_1. One can find that even though all of these sampling schemes have the same convergence rate of $O(N^{-1/2})$, the estimator based on Monte Carlo Almost Optimal (MAO) scheme [10], which samples the matrix elements according to their importance, yields significant smaller variance than the other two.

7.4.3 Case Study 3: Image Recovery

In this case study, we investigate the image processing technology of using a small number of pixel samples to recover an incomplete or fuzzy image. The strategy of Monte Carlo sampling plays a critical role for the quality of the recovered image.

We use an aerial image chosen from the USC-SIPI Image Database [11] as an example. The image recovery method is based on a matrix completion algorithm by optimizing the constrained nuclear norm of the matrix [12]. Two sets of pixel samples (12.19 % samples in each) are used—one is generated by uniform sampling

Table 7.1 Uniform, original, and MAO sampling scheme with different transition matrices

Sampling scheme	P_{ij}	P
Uniform sampling	$\dfrac{1}{n+1}$	$\begin{bmatrix} 0.25 & 0.25 & 0.25 \\ 0.25 & 0.25 & 0.25 \\ 0.25 & 0.25 & 0.25 \end{bmatrix}$
Original (Ulam and von Neumann)	$\lvert H_{ij} \rvert$	$\begin{bmatrix} 0.1 & 0.45 & 0.225 \\ 0.15 & 0.1 & 0.3 \\ 0.18 & 0.36 & 0.1 \end{bmatrix}$
MAO (Monte Carlo Almost Optimal)	$\dfrac{\lvert H_{ij} \rvert}{\sum_k \lvert H_{ik} \rvert}$	$\begin{bmatrix} 0.129032 & 0.580645 & 0.290323 \\ 0.272727 & 0.181818 & 0.545455 \\ 0.28125 & 0.5625 & 0.15625 \end{bmatrix}$

and the other by importance sampling. The importance sampling process consists of two stages:

(1) initial uniform sampling is performed with 3.05 % pixel samples to learn the rough pixel distribution in the image to produce a proposal function; and
(2) importance sampling is performed based on the proposal function created in (1) to generate the rest 9.14 % samples.

As shown in Fig. 7.4, the pixel samples generated by the importance sampling scheme lead to a recovered image in significantly higher quality than the one by using uniform samples.

7.4.4 Case Study 4: Matrix Multiplication

In this case study, we investigate the Monte Carlo methods of approximating the product of large matrices. Let A be an $m \times n$ matrix and B be an $n \times p$ matrix, where m, n, and p are large. The Monte Carlo sampling algorithm to fast approximate the product matrix $C = AB$ using s samples is described as follows [13]:

(1) Generate s random integers i_k between 1 and n with probability p_{i_k}, for $k = 1, \cdots, s$;
(2) Set $M^{(k)} = A^{(i_k)} / \sqrt{s p_{i_k}}$ and $N_{(k)} = B_{(i_k)} / \sqrt{s p_{(i_k)}}$, for $k = 1, \cdots, s$;
(3) Compute the matrix product of MN.

The fundamental idea of the Monte Carlo sampling algorithm is to construct a discrete random variable X with probability $p\left(X = \dfrac{A^{(k)} B_{(k)}}{p_k} \right) = p_k$ for $k = 1, \cdots, n$, where $A^{(k)}$ and $B_{(k)}$ represent the kth column of A and the kth row of B, respectively. The expectation of X is

Fig. 7.4 Comparison of image recovery from pixel samples generated by uniform sampling and importance sampling

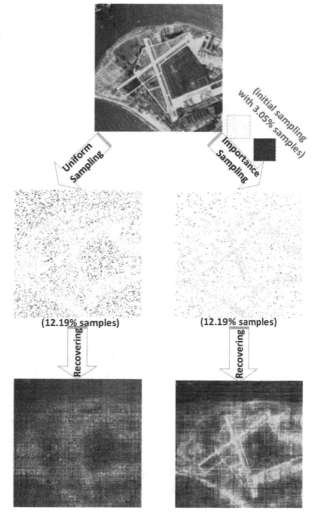

$$E(X) = \sum_{k=1}^{n} \frac{A^{(k)}B_{(k)}}{p_k} p_k = \sum_{k=1}^{n} A^{(k)}B_{(k)},$$

which is identical to matrix C. Therefore, matrix C can be fast approximated from the product of MN, where $MN = \frac{1}{s}\sum_{k=1}^{s} \frac{A^{(i_k)}B_{(i_k)}}{p_{i_k}}$ is an estimator. The variance on each element in MN is [13]

$$Var\left[(MN)_{ij}\right] = \sum_{k=1}^{n} \frac{A_{ik}^2 A_{kj}^2}{s^2 p_k} - \frac{1}{s^2}(AB)_{ij}^2.$$

To improve the accuracy of approximating matrix–matrix multiplication, importance sampling can be effectively applied by using the optimal probabilities

$$p_k = \frac{\left|A^{(k)}\right|\left|B_{(k)}\right|}{\sum_{j=1}^{n}\left|A^{(j)}\right|\left|B_{(j)}\right|},$$

where the expectation of approximation error can be theoretically minimized [13, 14].

Let us consider a simple toy example where

$$A = \begin{bmatrix} 1.6254 & 8.5799 & 8.5596 & 5.4832 & 9.1047 & 1.3797 \\ 9.8717 & 4.5759 & 7.5434 & 1.1230 & 8.4098 & 8.3874 \\ 4.8246 & 3.9563 & 5.1322 & 1.0143 & 8.6938 & 7.7045 \\ 0.9104 & 545.3701 & 4.6611 & 769.1278 & 9.1947 & 4.4734 \\ 14.9880 & 6.4549 & 197.7756 & 5.2276 & 88.9237 & 7.4417 \end{bmatrix}$$

and

$$B = \begin{bmatrix} 5.5899 & 2.6152 & 7.8102 & 8.2298 \\ 5.9020 & 666.2519 & 2.6396 & 7.7184 \\ 7.3150 & 143.9785 & 9.0632 & 4.9668 \\ 2.1493 & 695.7888 & 3.2657 & 974.7106 \\ 1.3544 & 758.2126 & 2.7377 & 4.6348 \\ 9.1003 & 8.3607 & 6.7709 & 27.5470 \end{bmatrix}$$

Figure 7.5 compares the relative approximation error of the resulting product matrix using uniform sampling and the one using important sampling with respect to sample size in 1000 runs. One can clearly find that when optimal selection probabilities are used, importance sampling outperforms uniform sampling with better approximation of the matrix product.

7.4.5 Case Study 5: Low-Rank Approximation

Given a matrix A, it is often desirable to find a good low-rank approximation to A in many data analysis applications. Denote $u_1, \cdots, u_n$ and $v_1, \cdots, v_n$ as the left and right singular vectors, respectively, $\sigma_1, \sigma_2, \cdots, \sigma_n$, are singular values in non-increasing order, the matrix A can be expressed as

$$A = \sum_{i=1}^{n} \sigma_i u_i v_i^T$$

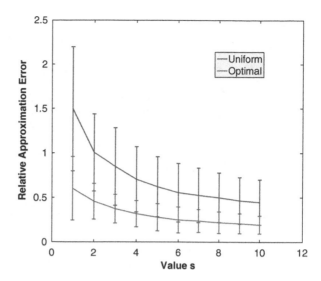

Fig. 7.5 Comparison of relative approximation error in matrix–matrix multiplication using uniform sampling and importance sampling with respect to sample size in 1000 Monte Carlo runs

The best k-rank approximation [15] to a matrix A is formed as $A_k = \sum_{i=1}^{k} \sigma_i u_i v_i^T$ with minor error

$$\min_{rank(B)=k} \|A - B\|_2 = \|A - A_k\|_2 = \sigma_{k+1}.$$

To handle the computational challenges involved in very big matrices, randomized Singular Value Decomposition (SVD) algorithm with Gaussian sampling [16–18] is widely used to approximate top-k singular values and singular vectors. The rationale is to construct a small condensed subspace by sampling A, where the dominant actions of A could be fast estimated from this small subspace with relatively low computation cost and high confidence. The procedure of randomized SVD is described as follows:

(1) Sample A with a standard Gaussian matrix Ω so that $Y = A\Omega$;
(2) Construct a basis Q for the range of Y;
(3) Compute matrix multiplication of $B = A^T Q$;
(4) Perform a deterministic SVD decomposition on $B = U_B \Sigma_B V_B^T$;
(5) Assign $u_j = Q v_{B_j}$, $\sigma_j = \sigma_{B_j}$, and $v_j = u_{B_j}$, $j = 1, \cdots, k$.

We consider an example of applying the randomized SVD algorithm to obtain a low-rank approximation of an image while minimizing the approximation error. The control variates approach is applied. First of all, the whole range space of matrix A is sampled to approximate the largest r singular components and then derive an approximate $\widetilde{A}_r$ such that $\widetilde{A}_r = \sum_{i=1}^{r} \sigma_i u_i v_i^T$. If the approximation error is too high, $\widetilde{A}_r$ is used as the control variate to sample the next dominating singular components. This process is repeated until satisfactory approximation accuracy is achieved.

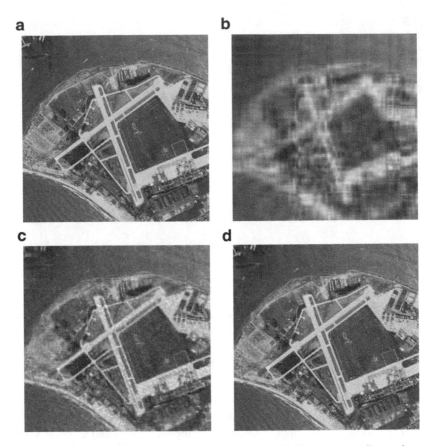

Fig. 7.6 Low-rank approximation using randomized SVD with Gaussian sampling and control variates strategy. (**a**) The original image. (**b**) Rank 10 with approximation error 14.3 %. (**c**) Rank 40 with approximation error 2.1 %. (**d**) Rank 100 with approximation error 0.9 %

Figure 7.6 shows the results of applying randomized SVD algorithm and control variate to compute a low-rank approximation of the aerial image with less than 1 % error. Figure 7.6a presents the original image. Figures 7.6b–d illustrate the adaptive reconstructed images with increasing numbers of singular components. Finally with 100 singular components, a low-rank approximation of the original image with 0.9 % error is obtained.

7.5 Summary

Monte Carlo methods are powerful tools to analyze large data sets, particularly in the big data era when the data growth outpaces the computer processing capability. We review the basics of Monte Carlo methods and the popular techniques for variance reduction in this article. Several application examples using Monte Carlo for data analysis are presented. The rule of thumb in Monte Carlo is, the more known knowledge is incorporated into the estimator, the more uncertainty can be reduced and the better data analysis accuracy can be obtained.

Acknowledgements This work is partially supported by NSF grant 1066471 for Yaohang Li and Hao Ji acknowledges support from ODU Modeling and Simulation Fellowship.

References

1. M. Hilbert, P. López, The world's technological capacity to store, communicate, and compute information. Science **332**, 60–65 (2011)
2. R.L. Burden, J.D. Faires, *Numerical Analysis* (Brooks/Cole, Cengage Learning, Boston, 2011)
3. J.M. Hammersley, D.C. Handscomb, *Monte Carlo Methods* (Chapman and Hall, Methuen & Co., London, and Wiley, New York, 1964)
4. J.S. Liu , *Monte Carlo Strategies in Scientific Computing* (Springer, New York, 2008)
5. Y. Li, M. Mascagni, Grid-based Monte Carlo application, in *Grid Computing Third International Workshop/Conference* (2002), pp. 13–24
6. Y. Li, M. Mascagni, Analysis of large-scale grid-based Monte Carlo applications. Int. J. High Perform. Comput. Appl. **17**, 369–382 (2003)
7. G.E. Forsythe, R.A. Leibler, Matrix inversion by a Monte Carlo method. Math. Comput. **4**, 127–129 (1950)
8. H. Ji, Y. Li , Reusing random walks in Monte Carlo methods for linear systems, in *Proceedings of International Conference on Computational Science*, vol. 9 (2012), pp. 383–392
9. H. Ji, M. Mascagni, Y. Li, Convergence analysis of Markov chain Monte Carlo linear solvers using Ulam–von Neumann Algorithm. SIAM J. Numer. Anal. **51**, 2107–2122 (2013)
10. I.T. Dimov, B. Philippe, A. Karaivanova, C. Weihrauch, Robustness and applicability of Markov chain Monte Carlo algorithms for eigenvalue problems. Appl. Math. Model. **32**, 1511–1529 (2008)
11. A. Weber, *SIPI image database*, The USC-SIPI Image Database, Signal & Image Processing Institute, Department of Electrical Engineering, Viterbi School of Engineering, Univ. of Southern California (2012), http://sipi.usc.edu/database
12. J.F. Cai, E.J. Candès, Z. Shen, A singular value thresholding algorithm for matrix completion. SIAM J. Optim. **20**, 1956–1982 (2010)
13. P. Drineas, R. Kannan, M.W. Mahoney, Fast Monte Carlo algorithms for matrices I: approximating matrix multiplication. SIAM J. Comput. **36**, 132–157 (2006)
14. S. Eriksson-Bique, M. Solbrig, M. Stefanelli, S. Warkentin, R. Abbey, I.C. Ipsen, Importance sampling for a Monte Carlo matrix multiplication algorithm, with application to information retrieval. SIAM J. Sci. Comput. **33**, 1689–1706 (2011)
15. G.H. Golub, C.F. Van Loan, *Matrix Computations* (Johns Hopkins University Press, Baltimore, 2012)
16. N. Halko, P.G. Martinsson, J.A. Tropp, Finding structure with randomness: Probabilistic algorithms for constructing approximate matrix decompositions. SIAM Rev. **53**, 217–288 (2011)
17. M.W. Mahoney, Randomized algorithms for matrices and data. Found. Trends Mach. Learn. **3**, 123–224 (2011)
18. H. Ji, Y. Li, GPU accelerated randomized singular value decomposition and its application in image compression, in *Proceedings of Modeling, Simulation, and Visualization Student Capstone Conference*, Sulfolk (2014)

Part III
Selected Topics in Data Science

Chapter 8
Feature Extraction via Vector Bundle Learning

Risheng Liu and Zhixun Su

Abstract In this paper, we propose a geometric framework, called Vector Bundle Learning (VBL) for feature extraction. In our framework, a vector bundle is considered as the intrinsic structure to extract features from high dimensional data. By defining a manifold to model the structure of sample set, features sampled from each fibre of a vector bundle can be obtained by metric learning on the manifold. A number of existing algorithms can be reformulated and explained in this unified framework. Based on the proposed framework, a novel supervised feature extraction algorithm called Vector Bundle Discriminant Analysis (VBDA) is proposed for recognition and classification. Experimental results on face recognition and handwriting digits classification demonstrate the excellent performance of our VBDA algorithm.

8.1 Introduction

Feature extraction is an extensively discussed topic in computer version and pattern recognition fields. Different kinds of techniques have been developed for this key problem. In this paper, we propose to view feature extraction from a perspective of vector bundle.

There have been many feature extraction algorithms which can be categorized as linear and nonlinear approaches. Traditional linear feature extraction algorithms include Principal Component Analysis (PCA) [1, 2], Multidimensional Scaling (MDS) [3] and Linear Discriminant Analysis (LDA) [4].

During the recent years, a number of nonlinear feature extraction algorithms called manifold learning have been developed. They seek to find the intrinsic low dimensional nonlinear structure, which is regarded as a manifold embedded in a high dimensional space. The most well-known unsupervised algorithms include Isomap [5], Diffusion Map [6], Laplacian Eigenmap (LE) [7], Local Linear Embedding (LLE) [8], Hessian Eigenmap (HE) [9] and Local Tangent Space Alignment (LTSA)

R. Liu • Z. Su (✉)
Dalian University of Technology, Dalian, China
e-mail: rsliu@dlut.edu.cn; zxsu@dlut.edu.cn

© Springer International Publishing Switzerland 2015
L.M. Chen et al., *Mathematical Problems in Data Science*,
DOI 10.1007/978-3-319-25127-1_8

[10], and the typical supervised algorithms include Maximum Margin Criterion (MMC) [11], Average Neighborhood Margin Maximization (ANMM) [12] and Semi-Riemannian Discriminant Analysis (SRDA) [13].

Recently, there is considerable interest in using linear methods, inspired by the geometric intuition of manifold learning, to find the nonlinear structure of features. The representative algorithms include Local Preserving Projection (LPP) [14, 15], Unsupervised Discriminant Projection (UDP) [16], Neighborhood Preserving Embedding (NPE) [17], Laplacian PCA (LPCA) [18], Marginal Fisher Analysis (MFA) [19] and Local Discriminant Embedding (LDE) [20].

In this paper, we propose a geometric framework, called Vector Bundle Learning (VBL), to provide a general theoretical analysis and a unified perspective on feature extraction. By geometrization of the sample structure, we get a so-called structural manifold, where the geometry of the sample set is prescribed by the metric tensor of this manifold. If define a Riemannian metric tensor on the structural manifold, we can obtain low dimensional approximations in each fibre by the local metric learning on the structural manifold. Then total low dimensional approximations can be obtained by aligning all fibres together. Many existing feature extraction algorithms can be reformulated by our VBL framework. Furthermore, a novel supervised feature extraction algorithm, called Vector Bundle Discriminant Analysis (VBDA), is proposed based on our VBL framework.

The rest of this paper is organized as follows. We first present our VBL framework in Sect. 8.2. The VBDA algorithm is presented in Sect. 8.3. The experimental results are shown in Sect. 8.4. Finally, we conclude the paper in Sect. 8.6.

8.2 Vector Bundle Learning

In this section, we first introduce two important geometric concepts, then present the Vector Bundle Learning (VBL) framework to provide a novel geometric perspective to understand and formulate feature extraction. The meanings of notations used hereafter can be found in Table 8.1.

8.2.1 Mathematical Fundamentals

Vector bundle as a special fibre bundle is a topological construction which makes precise the idea of a family of vector space parameterized by a manifold $\mathbb{M}$. Locally, the vector bundle is the tensor product of a neighbourhood on the base manifold and a d-dimensional Euclidean space. Therefore, a (smooth) rank d vector bundle $\{\mathbb{E}, \mathbb{M}, \pi\}$ [21] consists of a pair of smooth manifolds, $\mathbb{E}$ (the total space) and $\mathbb{M}$ (the base space), together with a surjective map $\pi : \mathbb{E} \to \mathbb{M}$ (the projection). For each $\mathbf{x} \in \mathbb{M}$, there is a d-dimensional vector space $\mathbb{E}_x$ (the fibre of $\mathbb{E}$) and a diffeomorphism $\pi^{-1} : \{\mathbf{x}\} \times \mathbb{R}^d \to \mathbb{E}_x$ (the local trivialization of $\mathbb{E}$).

Table 8.1 Mathematical notations

$tr(\mathbf{A})$	The trace of matrix $\mathbf{A}$	$\mathbf{A}^T$	The transpose of $\mathbf{A}$
$\mathbf{I}\,(\mathbf{I}_{p \times p})$	The identity matrix (of size $p \times p$)	$\mathbf{e}\,(\mathbf{e}_p)$	The all-one column vector
$\mathbf{D}_p$	$= [\mathbf{I}_{p \times p}, -\mathbf{e}_p]^T$	$\mathbf{H}_p$	$= \mathbf{I}_{p \times p} - \frac{1}{p}\mathbf{e}_p\mathbf{e}_p^T$
$\mathbb{E}_i$	The fibre over $\mathbf{x}_i$.	$\mathbb{E}$	$= \{\mathbb{E}_1, \ldots, \mathbb{E}_m\}$
$\mathbf{x}_i$	The ith sample in $\mathbb{R}^n$	$\mathscr{S}_x$	$= \{\mathbf{x}_1, \ldots, \mathbf{x}_m\}$
$\mathscr{S}_{x_i}$	$= \{\mathbf{x}_1^i, \ldots, \mathbf{x}_{m_i}^i, \mathbf{x}_i\}$	$\mathbf{X}$	$= [\mathbf{x}_1, \ldots, \mathbf{x}_m]$, an $n \times m$ matrix
$\mathbf{X}_i$	$= [\mathbf{x}_1^i, \ldots, \mathbf{x}_{m_i}^i, \mathbf{x}_i]$	$\mathbf{y}_i$	The approximation of $\mathbf{x}_i$ in $\mathbb{R}^d$
$\mathscr{S}_y$	$= \{\mathbf{y}_1, \ldots, \mathbf{y}_m\}$	$\mathscr{S}_{y_i}$	$= \{\mathbf{y}_1^i, \ldots, \mathbf{y}_{m_i}^i, \mathbf{y}_i\}$
$\mathbf{Y}$	$= [\mathbf{y}_1, \ldots, \mathbf{y}_m]$, a $d \times m$ matrix	$\mathbf{Y}_i$	$= [\mathbf{y}_1^i, \ldots, \mathbf{y}_{m_i}^i, \mathbf{y}_i]$
$d(\mathbf{y}_i, \mathbf{y}_j)$	The distance between $\mathbf{y}_i$ and $\mathbf{y}_j$	$\mathbf{U}$	The linear projection matrix
C_{y_i}	The class that $\mathbf{y}_i$ belongs to	$n(C_{y_i})$	The number of points in C_{y_i}

Another important concept in differential geometry is metric tensor. Suppose that $\mathbf{r}$ is a vector. Then a metric tensor $g(\mathbf{r}, \mathbf{r})$ is expressible as

$$g(\mathbf{r}, \mathbf{r}) = \mathbf{r}^T \boldsymbol{\Lambda} \mathbf{r}, \tag{8.1}$$

where $\boldsymbol{\Lambda}$ is a symmetric matrix. If the metric matrix $\boldsymbol{\Lambda}$ is positive definite at every point on the manifold, then g is called a Riemannian metric tensor. If $\boldsymbol{\Lambda}$ is only non-degenerate, then g is a semi-Riemannian metric tensor. One may refer to [22] for more details.

8.2.2 Vector Bundle Learning Framework

Suppose $\mathscr{S}_x$ is randomly sampled from a d-dimensional sub-manifold $\mathbb{M}$ which is embedded in an ambient Euclidean space $\mathbb{R}^n$. For each $\mathbf{x}_i \in \mathscr{S}_x$, there is a coordinate chart (V_{x_i}, φ_{x_i}) associated with it. We denote all the points sampled from V_{x_i} as $\mathscr{S}_{x_i}$ and call it the **reference sample set** of $\mathbf{x}_i$. Therefore, different from traditional manifold learning algorithms, we describe feature extraction as learning a trivialization for each fibre locally and aligning all fibres together globally.

In VBL framework, we use distance from $\mathbf{y}_i$ to other points in $\mathscr{S}_{y_i}$ to characterize the structural relationship of $\mathbb{E}_i$. So with $\mathbf{y}_i$ and $\mathscr{S}_{y_i}$, an m_i-tuple coordinate representation $\mathbf{d}_{y_i}$ is defined as

$$\mathbf{d}_{y_i} = [d(\mathbf{y}_i, \mathbf{y}_1^i), \ldots, d(\mathbf{y}_i, \mathbf{y}_{m_i}^i)]^T. \tag{8.2}$$

So we can say that $\mathbf{d}_{y_i}$ is a new point which contains the locally structural information of the original sample manifold. This $\mathbf{d}_{y_i}$ can also be considered as a point sampled from another manifold $\mathbb{S}$ which is called the **structural manifold** of $\mathbb{M}$ in VBL framework.

We use a Riemannian metric tensor g_R to furnish the newly built structural manifold, then features can be extracted by learning the Riemannian geometry of $\mathbb{S}$. Suppose the Riemannian metric matrix $\mathbf{G}_i^R$ at $\mathbf{x}_i$ has been determined. It is easy to see that $g_R(\mathbf{d}_{y_i}, \mathbf{d}_{y_i})$ can be written as

$$g_R(\mathbf{d}_{y_i}, \mathbf{d}_{y_i}) = \mathbf{d}_{y_i}^T \mathbf{G}_i^R \mathbf{d}_{y_i} = tr(\mathbf{Y}_i \mathbf{M}_i \mathbf{Y}_i^T), \tag{8.3}$$

where $\mathbf{M}_i = \mathbf{D}_{m_i} \mathbf{G}_i^R \mathbf{D}_{m_i}^T$. We call $\mathbf{M}_i$ the *fibre matrix* of $\mathbb{E}_i$. Assuming $\mathbf{M}_i$ has been determined, it is easy to confirm that $\mathbf{G}_i^R$ can also be uniquely determined as:

$$\mathbf{G}_i^R = \mathbf{E}_{m_i} \mathbf{M}_i \mathbf{E}_{m_i}^T, \tag{8.4}$$

where $\mathbf{E}_{m_i} = [\mathbf{I}_{m_i \times m_i}, \mathbf{0}_{m_i \times 1}] - [\mathbf{e}_{(m_i+1) \times 1}, \mathbf{0}_{(m_i+1) \times (m_i-1)}]^T$. Thus the optimization problem of learning fibre $\mathbb{E}_i$ can be defined as minimizing the values $g_R(\mathbf{d}_i, \mathbf{d}_i)$ of metric tensor, which is determined by the fibre matrix $\mathbf{M}_i$.

After performing manipulations on each fibre, the geometry of global vector bundle can be determined. If we aim at deriving the global approximation $\mathbf{Y}$ from $\mathbb{E}$, we should align all fibres together. This alignment skill has been used in [10] and extended in [18].

By defining an $m \times (m_i + 1)$ binary selection matrix $\mathbf{B}_{m_i}$ associated with the ith fibre whose structure is that $(\mathbf{B}_{m_i})_{pq} = 1$ if the qth vector in $\mathbf{Y}_i$ is the pth vector in $\mathbf{Y}$, the sum of metric tensor values can be written as:

$$\sum_{i=1}^{m} g_R(\mathbf{d}_{y_i}, \mathbf{d}_{y_i}) = tr(\mathbf{Y} \mathbf{M} \mathbf{Y}^T), \tag{8.5}$$

where $\mathbf{M} = \sum_{i=1}^{m} \mathbf{B}_{m_i} \mathbf{M}_i \mathbf{B}_{m_i}^T$.

So the global low dimensional approximation $\mathbf{Y}$ can be learnt via the following optimization problem:

$$\begin{cases} \arg\min_{\mathbf{Y}} tr(\mathbf{Y} \mathbf{M} \mathbf{Y}^T), \\ s.t. \ \mathbf{Y} \mathbf{C} \mathbf{Y}^T = \mathbf{I}, \end{cases} \tag{8.6}$$

where the constraint $\mathbf{Y} \mathbf{C} \mathbf{Y}^T = \mathbf{I}$ removes an arbitrary scaling factor. The constraint matrix $\mathbf{C}$ is dependent on applications so that it induces features with some invariant properties. It is typically a symmetric matrix and may also be the metric matrix of another structural manifold.[1] We will show in Sect. 8.2.3 that many existing algorithms can be reformulated by our VBL framework.

---

[1]It will be discussed in Sect. 8.3.

The solution of (8.6) can be obtained by solving the following generalized eigen-value problem [7, 8, 10]:

$$\mathbf{M}\mathbf{y}_i = \lambda_i \mathbf{C}\mathbf{y}_i, \qquad (8.7)$$

where λ_i are the 2nd to the $(d+1)$th smallest generalized eigen-values of the above problem.

Linearization By assuming that the low dimensional approximation can be obtained from linear projection: $\mathbf{Y} = \mathbf{U}^T\mathbf{X}$, the optimal projection matrix $\mathbf{U}$ can be found by solving the following problem:

$$\begin{cases} \arg\min_{\mathbf{U}} tr(\mathbf{U}^T\mathbf{X}\mathbf{M}\mathbf{X}^T\mathbf{U}), \\ s.t. \ \mathbf{U}^T\mathbf{X}\mathbf{C}\mathbf{X}^T\mathbf{U} = \mathbf{I}. \end{cases} \qquad (8.8)$$

Extensions A global view and semi-Riemannian generalization can also be proposed as extensions to this framework. For global algorithms, we choose all points in S_x to construct the structural manifold and define a global Riemannian metric tensor $\mathbf{G}^R$ for it. In this case, VBL can be considered as a global algorithm. For some algorithms (e.g., MMC, ANMM, SRDA), the positive-definiteness of the metric matrices is not fulfilled. So semi-Riemannian metric tensors should be used to generalize VBL. Henceforth, we use g with respect to a non-degenerate symmetric matrix $\mathbf{G}_i$ as the metric tensor in the generalized VBL framework.

8.2.3 General Framework for Feature Extraction

Yan et al. [19] proposed the Graph Embedding (GE) framework to unify feature extraction algorithms. In this section, we present a unified geometric perspective to understand feature extraction using our VBL, which can be considered as a more general framework.

Theorem 8.1. *By defining the metric matrix as:*

$$(\mathbf{G}_i^{GE})_{pq} = \begin{cases} \mathbf{W}_{ip}^{GE}, & if \ p = q \ and \ \mathbf{x}_p^i \in \mathscr{S}_{x_i}, \\ 0, & otherwise, \end{cases}$$

where $\mathbf{W}_{ip}^{GE}$ is the element of the similarity matrix in [19], we can get GE from our VBL framework.

With Theorem 8.1, it is easy to see that GE framework can only formulate algorithms which have a diagonal metric matrix, such as PCA, LDA, Isomap, LLE and LE/LPP. The diagonal constraint for metric matrix is so strong that lots of algorithms cannot be reformulated in GE. In contrast, our VBL relaxes the constraint "positive-definite diagonal" to "non-degenerate symmetric" and hance supports a more general metric matrix. Therefore, feature extraction algorithms mentioned in Sect. 8.1 which cannot be incorporated in GE, such as LTSA, UDP, HE, LPCA, Diffusion Map and discrepancy criterion based algorithms, e.g., MMC, ANMM and SRDA can all fit in our VBL framework. The difference among them lies in the ways to construct the reference sample set, the fibre matrix (or the metric matrix) and the constraint matrix.[2]

8.3 Vector Bundle Discriminant Analysis

VBL can be utilized as a general platform to develop new algorithms for feature extraction. In this section, as an example we introduce a novel supervised feature extraction algorithm, called Vector Bundle Discriminant Analysis (VBDA), based on the VBL framework.

By assuming that points in C_{y_i} are sampled from one fibre $\mathbb{E}_i^c$, the proposed VBDA aims to enhance both the local fibre similarity and the non-local vector bundle separability to optimize the intrinsic structure of group of classes for classification.

To measure the local fibre similarity, we model the intra-fibre structural manifold $\hat{\mathbb{S}}$ as:

$$\hat{\mathbf{d}}_{y_i} = [d(\mathbf{y}_i, \mathbf{y}_1^i), \dots, d(\mathbf{y}_i, \mathbf{y}_{\hat{m}_i}^i)]^T, \tag{8.9}$$

where $\mathbf{y}_j^i \in C_{y_i}$ and $\mathbf{y}_j^i \neq \mathbf{y}_i$ ($\hat{m}_i = n(C_{y_i}) - 1$).

With the newly built $\hat{\mathbb{S}}$, the Riemannian metric tensor $\hat{g}_R(\hat{\mathbf{d}}_{y_i}, \hat{\mathbf{d}}_{y_i})$ can be obtained as:

$$\hat{g}_R(\hat{\mathbf{d}}_{y_i}, \hat{\mathbf{d}}_{y_i}) = \hat{\mathbf{d}}_{y_i}^T \hat{\mathbf{G}}_i^R \hat{\mathbf{d}}_{y_i} = tr(\mathbf{Y}_i \mathbf{M}_i \mathbf{Y}_i^T), \tag{8.10}$$

where $\mathbf{M}_i = \mathbf{D}_{\hat{m}_i} \hat{\mathbf{G}}_i^R \mathbf{D}_{\hat{m}_i}^T$ is the fibre matrix of $\mathbb{E}_i^c$ and $\hat{\mathbf{G}}_i^R$ is constrained to be a diagonal matrix. The local similarity of each $\mathbb{E}_i^c$ can be learnt from the Riemannian geometry of $\hat{\mathbb{S}}$. Thus the problem reduces to finding an ambient Riemannian metric matrix $\hat{\mathbf{G}}_i^R$ at each $\hat{\mathbf{d}}_{y_i}$.

The dissimilarity of points (i.e., the Euclidean distance) in different fibres can be used to measure the non-local separability of vector bundle. So to enhance the inter-fibre separability, we model inter-fibre structural manifold $\check{\mathbb{S}}$ as

$$\check{\mathbf{d}}_{y_i} = [d(\mathbf{y}_i, \mathbf{y}_1^i), \dots, d(\mathbf{y}_i, \mathbf{y}_{\check{m}_i}^i)]^T, \tag{8.11}$$

where $\mathbf{y}_j^i \notin C_{y_i}$ ($\check{m}_i = m - n(C_{y_i})$).

---

[2]The details are given in the supplementary material due to the page limit.

Similar to the intra-fibre case, the Riemannian metric tensor $\check{g}_R(\check{\mathbf{d}}_{y_i}, \check{\mathbf{d}}_{y_i})$ for point on $\check{\mathbb{S}}$ can be obtained as

$$\check{g}_R(\check{\mathbf{d}}_{y_i}, \check{\mathbf{d}}_{y_i}) = \check{\mathbf{d}}_{y_i}^T \check{\mathbf{G}}_i^R \check{\mathbf{d}}_{y_i} = tr(\mathbf{Y}_i \mathbf{C}_i \mathbf{Y}_i^T), \tag{8.12}$$

where $\mathbf{C}_i = \mathbf{D}_{\check{m}_i} \check{\mathbf{G}}_i^R \mathbf{D}_{\check{m}_i}^T$ and $\check{\mathbf{G}}_i^R$ is also constrained to diagonal. The non-local separability of different fibres (vector bundle) can be learnt from another Riemannian metric matrix $\check{\mathbf{G}}_i^R$.

Different from LLE, which uses reconstruction weights, VBDA transfers the local geometry from the sample space to the feature space by learning metric in $\mathscr{S}_x$ and applying it to $\mathscr{S}_y$. Therefore, the smaller intra-fibre margins and larger inter-fibre margins in $\mathbb{E}$ both can be fulfilled by learning $\hat{\mathbf{G}}_i^R$ and $\check{\mathbf{G}}_i^R$.

In this paper, we introduce a method to learn the Riemannian metric matrix in the sample space. The ambient Riemannian metric tensor at $\mathbf{x}_i$ can be written as:

$$g_R(\mathbf{d}_{x_i}, \mathbf{d}_{x_i}) = \mathbf{d}_{x_i}^T \mathbf{G}_i^R \mathbf{d}_{x_i} = \mathbf{g}_i^T \mathbf{D}_{x_i} \mathbf{g}_i,$$

where

$$\mathbf{g}_i = [\sqrt{\mathbf{G}_i^R(1, 1)}, \dots, \sqrt{\mathbf{G}_i^R(m_i, m_i)}]^T$$

and

$$\mathbf{D}_{x_i} = \text{diag}(d(\mathbf{x}_i, \mathbf{x}_1^i)^2, \dots, d(\mathbf{x}_i, \mathbf{x}_{m_i}^i)^2).$$

Therefore, we can minimize this metric and obtain the following optimization problem to learn the intrinsic fibre structure:

$$\begin{cases} \arg\min_{\mathbf{g}_i} \mathbf{g}_i^T \mathbf{D}_{x_i} \mathbf{g}_i, \\ s.t. \ \mathbf{e}^T \mathbf{g}_i = 1. \end{cases} \tag{8.13}$$

It is easy to check that the solution to the above problem is

$$\mathbf{g}_i = \frac{(\mathbf{D}_{x_i})^{-1} \mathbf{e}}{\mathbf{e}^T (\mathbf{D}_{x_i})^{-1} \mathbf{e}}. \tag{8.14}$$

By applying the above model to intra-fibre and inter-fibre structural manifolds, the metric matrices $\hat{\mathbf{G}}_i^R$ and $\check{\mathbf{G}}_i^R$ can be obtained as

$$(\hat{\mathbf{G}}_i^R)_{pq} = \begin{cases} (\hat{\mathbf{g}}_i(p))^{\hat{r}}, & \text{if } p = q \\ 0, & \text{otherwise} \end{cases} \tag{8.15}$$

and

$$(\check{\mathbf{G}}_i^R)_{pq} = \begin{cases} (\check{\mathbf{g}}_i(p))^{\check{t}}, & \text{if } p = q \\ 0, & \text{otherwise} \end{cases}, \tag{8.16}$$

respectively. We empirically find that the discriminability will be enhanced if we choose $\hat{t}, \check{t} \in [0, 3]$ properly.

8.4 Experimental Results

Experiments are conducted on face recognition and handwriting digits classification to test the performance of VBDA against the existing algorithms such as LPP, UDP, LDA and MFA on recognition and classification.

8.4.1 Face Recognition

Face recognition experiments are conducted on a subset of facial data in experiment 4 of FRGC version 2 [23]. There are 8014 images of 466 subjects in the query set. We choose the first 30 images of each subject in this set if the number of images is not less than 30. Thus we get 2850 images of 95 subjects for our experiment. These images are all cropped to a size of 32×32. Figure 8.1 shows some cropped facial images in experiment 4 of FRGC version 2.

Note that the data are preprocessed by PCA. We follow the idea of Yan et al. [19] to choose the percentage of energy retained in PCA preprocessing step between 90–100 % in our experiment, respectively. The nearest neighbour classifier is used for classification with the extracted features. The image set of each subject is split into different gallery and probe sets, where Gm/Pn means m images are randomly selected for training and the remaining n images are for testing. Such a trial is repeated 20 times.

Fig. 8.1 Some cropped FRGC version 2 facial images

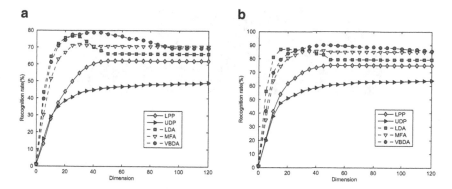

Fig. 8.2 The recognition rate curves versus the variation of dimensions on experiment 4 of FRGC version 2 database. (**a**) G3/P27; (**b**) G5/P25

Table 8.2 The average experimental results on experiment 4 of FRGC version 2 facial data and USPS handwriting digital data (mean ± std %). The best results are shown in bold font.

Method	FRGC facial data		USPS digital data	
	G3/P27	G5/P25	G30/P170	G40/P160
LPP	59.1 ± 4.0 (50, 93)	70.1 ± 2.2 (50, 92)	7 6.3 ± 7.6 (25)	87.6 ± 10.5 (20)
UDP	48.3 ± 4.3 (120, 99)	58.6 ± 3.5 (120, 98)	79.3 ± 6.5 (25)	88.8 ± 2.4 (20)
LDA	75.6 ± 4.8 (25, 93)	85.1 ± 2.2 (20, 92)	81.1 ± 7.3 (5)	89.0 ± 8.0 (15)
MFA	71.9 ± 4.5 (30, 91)	82.5 ± 2.4 (45, 92)	80.8 ± 2.9 (5)	87.6 ± 3.0 (10)
VBDA	**77.2 ± 4.1 (45, 97)**	**87.6 ± 2.1 (45, 99)**	**87.0 ± 3.2 (10)**	**93.3 ± 4.2 (25)**

The first number in the brackets is the optimal dimension of feature space and the second one is the percentage of energy retained in the PCA preprocessing step (for face recognition experiment)

Figure 8.2 displays the recognition rate curves versus the feature space dimensions when performing these methods. Table 8.2 shows the average recognition results on FRGC version 2. One can see that VBDA is better than other methods in comparison.

8.4.2   Handwriting Digits Classification

The USPS handwriting digital data[3] includes 10 classes from "0" to "9". Each class has 1100 samples. Images are cropped to 16 × 16. Figure 8.3 shows subsets of the cropped images of two digits. The first 200 images of each class are chosen for our experiment. We directly apply all algorithms to the normalized data without using PCA as preprocessing.

---

[3] Available at http://www.cs.toronto.edu/~roweis/data.html.

Fig. 8.3 Some cropped USPS digital images

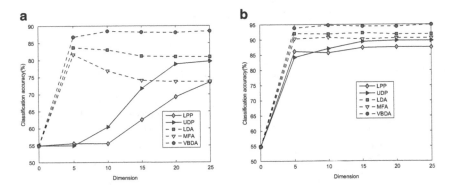

Fig. 8.4 The classification accuracy curves versus the variation of dimensions on USPS database. (**a**) G30/P170; (**b**) G40/P160

Figure 8.4 illustrates the classification accuracy curves versus the variation of dimensions. The average classification results are also shown in Table 8.2. The performance of VBDA is significantly better than other methods under consideration.

8.4.3 Discussions

The free parameters in the test methods were determined in the following way: as in LPP, UDP and MFA, the number of reference sample set (k, $\hat{k}$ and $\check{k}$) is selected according to the original data in different experiments. For face recognition, the parameter k in LPP and UDP is set to 5, the parameters $\hat{k}$ and $\check{k}$ in MFA are set to $m - 1$ (m is the number of gallery set) and 210, respectively. For handwriting digits classification, we set the parameter k to 5 in LPP and UDP, the parameters $\hat{k}$ and $\check{k}$ to 7 and 40, respectively, in MFA. The Gaussian kernel $\exp\left(-\frac{\|x_i - x_j\|^2}{t}\right)$ is used in LPP and UDP. We set the parameter t to 250 in LPP and 200 in UDP, respectively.

8.5 Unifying Previous Algorithms in VBL Framework

8.5.1 Local Algorithms

Local algorithms aim at preserving the structure for local fibres. So in these algorithms, we choose $\mathbf{x}_i$ and its k nearest neighbours as $\mathscr{S}_{x_i}$.[4]

LLE [8] defines the fibre matrix as $\mathbf{M}_i = \mathbf{A}_i^T \mathbf{A}_i$, where $\mathbf{A}_i = [1, -\mathbf{w}_1^i, \ldots, -\mathbf{w}_{m_i}^i]$ are the local representation coefficients. Let $\omega_i = [\mathbf{w}_1^i, \mathbf{w}_2^i, \ldots, \mathbf{w}_{m_i}^i]^T$ and the local Gram matrix Γ_i be $(\Gamma_i)_{pq} = (\mathbf{x}_i - \mathbf{x}_p^i)^T (\mathbf{x}_i - \mathbf{x}_q^i)$, one can calculate ω_i like this:

$$\begin{cases} \omega_i = (\Gamma_i^{-1}\mathbf{e})^T \\ \omega_i \leftarrow \omega_i/\omega_i\mathbf{e} \end{cases}.$$

NPE [17] can be considered as a linearization of LLE.

If we view the local tangent space as the fibre at every point and define $\mathbf{M}_i = \mathbf{H}_{m_i+1} - \mathbf{Q}_i\mathbf{Q}_i^T$, where $\mathbf{Q}_i$ is the matrix consisting of the eigen-vectors associated with the d largest eigen-values of $\mathbf{X}_i^T\mathbf{X}_i$, LTSA [10] can also be fit into our VBL framework.

To fit LE [7] into VBL, we may define the metric matrix as

$$(\mathbf{G}_i^{LE})_{pq} = \begin{cases} \exp\left(-\dfrac{\|\mathbf{x}_i - \mathbf{x}_p^i\|^2}{t}\right), & \text{if } p = q \text{ and } \mathbf{x}_p^i \in \mathscr{S}_{x_i}, \\ 0, & \text{otherwise}, \end{cases}$$

then the fibre matrix can be given as $\mathbf{M}_i = \mathbf{D}_{m_i}\mathbf{G}_i^{LE}\mathbf{D}_{m_i}^T$. The constraint matrix $\mathbf{C} = \mathbf{D}^{LE}$, where $\mathbf{D}^{LE}$ is a diagonal matrix whose entries are sums of $\mathbf{G}_i^{LE}$. LPP [14] is the linearization of LE.

Let the fibre matrix $\mathbf{M}_i$ be the same as the one in LPP, and the constraint matrix $\mathbf{C} = \mathbf{H}_m$, UDP [16] can also be fit into our VBL framework.

HE [9] can be viewed as a variant of LE where the Hessian is in place of the Laplacian. The fibre matrix can be obtained as $\mathbf{M}_i = (\mathbf{H}_i^{tan})^T\mathbf{H}_i^{tan}$, where $\mathbf{H}_i^{tan}$ is the tangent Hessian matrix.

The fibre matrix of LPCA [18] is exactly the local Laplacian scatter matrix:

$$\mathbf{M}_i = \mathbf{W}_i - \frac{\mathbf{W}_i\mathbf{e}\mathbf{e}^T\mathbf{W}_i}{\mathbf{e}^T\mathbf{W}_i\mathbf{e}},$$

where $\mathbf{W}_i$ is determined by coding length.

MFA [19] and LDE [20] are essentially the same. In VBL, we choose $\mathscr{S}_{x_i} = \{\mathbf{x}_i\} \cup \mathscr{N}_{\hat{k}}(\mathbf{x}_i) \cup \mathscr{N}_{\check{k}}(\mathbf{x}_i)$, where $\mathscr{N}_{\hat{k}}(\mathbf{x}_i)$ is the $\hat{k}$ nearest within-class neighbours and

---

[4]For MFA, LDE, ANMM and SRDA, the choice of $\mathscr{S}_{x_i}$ uses the class information as they are supervised dimensionality reduction techniques. See the details that follow.

$\check{\mathcal{N}}_{\check{k}}(\mathbf{x}_i)$ is the $\check{k}$ nearest between-class neighbours, and use $\hat{\mathcal{N}}_{\hat{k}}(\mathbf{x}_i)$ to construct the metric matrix and $\check{\mathcal{N}}_{\check{k}}(\mathbf{x}_i)$ to construct the constraint matrix. First we define

$$(\mathbf{G}_i^{MFA})_{pq} = \begin{cases} 1, & \text{if } p = q \text{ and } \mathbf{x}_p^i \in \hat{\mathcal{N}}_{\hat{k}}(\mathbf{x}_i), \\ 0, & \text{otherwise.} \end{cases}$$

Then the fibre matrix is given as $\mathbf{M}_i = \mathbf{D}_{m_i}\mathbf{G}_i^{MFA}\mathbf{D}_{m_i}^T$. Next, let $\mathbf{W}^{MFA}$ be an $m \times m$ symmetric matrix with

$$(\mathbf{W}^{MFA})_{pq} = \begin{cases} 1, & \text{if } \mathbf{x}_p \in \check{\mathcal{N}}_{\check{k}}(\mathbf{x}_q) \text{ or } \mathbf{x}_q \in \check{\mathcal{N}}_{\check{k}}(\mathbf{x}_p), \\ 0, & \text{otherwise,} \end{cases}$$

and let $\mathbf{D}^{MFA}$ be a diagonal matrix whose entries are column sums of $\mathbf{W}^{MFA}$. Then the constraint matrix $\mathbf{C} = \mathbf{L}^{MFA}$, where $\mathbf{L}^{MFA} = \mathbf{D}^{MFA} - \mathbf{W}^{MFA}$. Thus we can get MFA and LDE from VBL.

ANMM [12] uses the average neighbourhood margin to find an embedding space that characterizes the within-class similarity and the between-class separability. We may choose k nearest within-class neighbours and k nearest between-class neighbours as $\mathscr{S}_{x_i}$. Then

$$\mathbf{G}_i^{ANMM} = \begin{bmatrix} \hat{\mathbf{G}}_{\hat{k} \times \hat{k}} & 0 \\ 0 & -\check{\mathbf{G}}_{\check{k} \times \check{k}} \end{bmatrix},$$

where $\hat{\mathbf{G}}_{\hat{k} \times \hat{k}} = \frac{1}{\hat{k}}\mathbf{I}_{\hat{k} \times \hat{k}}$ and $\check{\mathbf{G}}_{\check{k} \times \check{k}} = \frac{1}{\check{k}}\mathbf{I}_{\check{k} \times \check{k}}$. And $\mathbf{M}_i = \mathbf{D}_{m_i}\mathbf{G}_i^{ANMM}\mathbf{D}_{m_i}^T$.

In SRDA [13], the structure of classes is conceptualized as a semi-Riemannian manifold and semi-Riemannian manifold learning is used to acquire the best embedding space for classification. Let $\mathscr{S}_{x_i} = \overset{K}{\underset{j=0}{\cup}} \mathcal{N}_{k_j}(\mathbf{x}_i)$ be the reference sample set, where $\mathcal{N}_{k_0}(\mathbf{x}_i)$ is the set of $\mathbf{x}_i$'s k_0 nearest within-class neighbours and $\mathcal{N}_{k_j}(\mathbf{x}_i)$ is the set of $\mathbf{x}_i$'s k_j nearest neighbours in the jth nearest class ($j = 1, \ldots, K$) [13]. Then $\mathbf{M}_i = \mathbf{D}_{m_i}\mathbf{G}_i^{SRDA}\mathbf{D}_{m_i}^T$, where $\mathbf{G}_i^{SRDA}$ is a semi-Riemannian metric matrix which is learnt by optimization.

8.5.2 Global Algorithms

In contrast with local techniques, global algorithms seek to directly find the intrinsic global structure hidden in the ambient space. For global algorithms, $\mathscr{S}_x$ is considered as the reference sample set. The fibre matrix $\mathbf{M}$ (or the metric matrix $\mathbf{G}$) can be defined directly from $\mathscr{S}_x$.

PCA [1] is by far the most popular unsupervised global linear algorithm for feature extraction. We can fit it into VBL by defining $\mathbf{M} = -\mathbf{H}_m$.

MDS [3] finds an embedding space that retains the pairwise Euclidean distance at the best. Let $\mathbf{M} = -\tau(\mathbf{D}^e)$, where $\tau(\mathbf{D}^e)$ is the inner product matrix corresponding to the Euclidean distance matrix $\mathbf{D}^e$.

Isomap [5] shares the same perspective with MDS but retains pairwise geodesic distance at the best. We may define the bundle matrix as $\mathbf{M} = -\tau(\mathbf{D}^g)$, where $\mathbf{D}^g$ is the geodesic distance matrix.

Diffusion Map [6] is based on defining a Markov random walk on the graph and a so-called diffusion distance to measure the similarity between features. A matrix $\mathbf{P}^{(1)}$ is formed with entries $\mathbf{P}_{ij}^{(1)} = K_{ij}/\sum_k K_{ik}$, where $K_{ij} = \exp\left(-\frac{\|x_i - x_j\|^2}{t}\right)$. Then we can fit it into VBL framework by defining $\mathbf{M} = -(\mathbf{P}^{(1)})^t$.

LDA [4] is well known as a classic technique in discriminant analysis. While PCA seeks to find a subspace that is efficient for representation, LDA seeks a subspace that is efficient for discrimination. Let $\mathbf{W}^{Class}$ be an $m \times m$ symmetric matrix with

$$\mathbf{W}_{pq}^{Class} = \begin{cases} \frac{1}{n(C_i)}, & \text{if } \mathbf{x}_p, \mathbf{x}_q \in C_i, \\ 0, & \text{otherwise,} \end{cases}$$

where C_i is the ith class. The within-class scatter and the between-class scatter matrices can be rewritten as:

$$\mathbf{S}^w = \sum_{i=1}^c \sum_{j=1}^{n(C_i)} (\mathbf{x}_j^i - \bar{\mathbf{x}}_i)(\mathbf{x}_j^i - \bar{\mathbf{x}}_i)^T \tag{8.17}$$

$$= \mathbf{X}(\mathbf{I} - \mathbf{W}^{Class})\mathbf{X}^T \tag{8.18}$$

and

$$\mathbf{S}^b = \sum_{i=1}^c n(C_i)(\bar{\mathbf{x}}_i - \bar{\mathbf{x}})(\bar{\mathbf{x}}_i - \bar{\mathbf{x}})^T \tag{8.19}$$

$$= \mathbf{X}(\mathbf{W}^{Class} - \frac{1}{m}\mathbf{e}\mathbf{e}^T)\mathbf{X}^T, \tag{8.20}$$

respectively, where $\bar{\mathbf{x}}_i = \frac{1}{n(C_i)}\mathbf{X}_i\mathbf{e}$ is the center of C_i and $\bar{\mathbf{x}} = \frac{1}{m}\mathbf{X}\mathbf{e}$ is the center of $\mathcal{S}_x$. Then letting $\mathbf{M} = \mathbf{I} - \mathbf{W}^{Class}$ and $\mathbf{C} = \mathbf{W}^{Class} - \frac{1}{m}\mathbf{e}\mathbf{e}^T$, LDA is fit into VBL framework.

MMC [11] takes the role of integrating within-class scatters and between-class scatters by differences, instead of ratios used by the traditional LDA. Because $\mathbf{S}^w - \mathbf{S}^b = \mathbf{X}(\mathbf{I} + \frac{1}{m}\mathbf{e}\mathbf{e}^T - 2\mathbf{W}^{Class})\mathbf{X}^T$, we have $\mathbf{M} = \mathbf{I} + \frac{1}{m}\mathbf{e}\mathbf{e}^T - 2\mathbf{W}^{Class}$ for MMC in VBL framework.

Table 8.3 lists the fibre matrices $\mathbf{M}_i$ (or the bundle matrix $\mathbf{M}$) and the constraint matrices $\mathbf{C}$ for all the above-mentioned methods.

Table 8.3 Different algorithms expressed in our VBL Framework

Algorithm	$\mathbf{M}_i$ or $\mathbf{M}$	$\mathbf{C}$	Type
LLE/NPE	$\mathbf{M}_i = \mathbf{A}_i^T \mathbf{A}_i$	$\mathbf{C} = \mathbf{I}$	N/L
LTSA	$\mathbf{M}_i = \mathbf{H}_{m_i+1} - \mathbf{Q}_i \mathbf{Q}_i^T$	$\mathbf{C} = \mathbf{I}$	N
LE/LPP	$\mathbf{M}_i = \mathbf{D}_{m_i} \mathbf{G}_i^{LE} \mathbf{D}_{m_i}^T$	$\mathbf{C} = \mathbf{D}^{LE}$	N/L
UDP	$\mathbf{M}_i = \mathbf{D}_{m_i} \mathbf{G}_i^{LE} \mathbf{D}_{m_i}^T$	$\mathbf{C} = \mathbf{H}_m$	L
HE	$\mathbf{M}_i = (\mathbf{H}_i^{tan})^T \mathbf{H}_i^{tan}$	$\mathbf{C} = \mathbf{I}$	N
LPCA	$\mathbf{M}_i = \mathbf{W}_i - \frac{\mathbf{W}_i \mathbf{e} \mathbf{e}^T \mathbf{W}_i}{\mathbf{e}^T \mathbf{W}_i \mathbf{e}}$	$\mathbf{C} = \mathbf{X}^\dagger (\mathbf{X}^\dagger)^T$	L
MFA(LDE)	$\mathbf{M}_i = \mathbf{D}_{m_i} \mathbf{G}_i^{MFA} \mathbf{D}_{m_i}^T$	$\mathbf{C} = \mathbf{L}^{MFA}$	L
ANMM	$\mathbf{M}_i = \mathbf{D}_{m_i} \mathbf{G}_i^{ANMM} \mathbf{D}_{m_i}^T$	$\mathbf{C} = \mathbf{X}^\dagger (\mathbf{X}^\dagger)^T$	L
SRDA	$\mathbf{M}_i = \mathbf{D}_{m_i} \mathbf{G}_i^{SRDA} \mathbf{D}_{m_i}^T$	$\mathbf{C} = \mathbf{X}^\dagger (\mathbf{X}^\dagger)^T$	L
PCA	$\mathbf{M} = -\mathbf{H}_m$	$\mathbf{C} = \mathbf{X}^\dagger (\mathbf{X}^\dagger)^T$	L
MDS	$\mathbf{M} = -\tau(\mathbf{D}_e)$	$\mathbf{C} = \mathbf{I}$	L
Isomap	$\mathbf{M} = -\tau(\mathbf{D}_g)$	$\mathbf{C} = \mathbf{I}$	N
Diffusion Map	$\mathbf{M} = -(\mathbf{P}^{(1)})^t$	$\mathbf{C} = \mathbf{I}$	N
LDA	$\mathbf{M} = \mathbf{I} - \mathbf{W}^{Class}$	$\mathbf{C} = \mathbf{W}^{Class} - \frac{1}{m} \mathbf{e} \mathbf{e}^T$	L
MMC	$\mathbf{M} = \mathbf{I} + \frac{1}{m} \mathbf{e} \mathbf{e}^T - 2\mathbf{W}^{Class}$	$\mathbf{C} = \mathbf{X}^\dagger (\mathbf{X}^\dagger)^T$	L

N and L means nonlinear VBL and the linearization of VBL. $\mathbf{X}^\dagger$ means the Moore–Penrose inverse of $\mathbf{X}$

8.6 Conclusions

We have proposed a general framework, called VBL, to unify the existing feature extraction techniques and develop new algorithms. Most traditional algorithms can be reformulated and understood in this framework. A novel supervised feature extraction algorithm, called VBDA, is also presented based on the proposed framework, by learning special Riemannian metric tensors. The experimental results have shown that our proposed VBDA is promising.

Acknowledgment We thank Prof. Zhouchen Lin for valuable comments and suggestions for improving this work. This work is partly supported by National Natural Science Foundation of China (Nos. 61300086, 61173103, 61432003, 6157209) and Fundamental Research Funds for the Central Universities (No. DUT15QY15).

References

1. I. Jolliffe, *Principal Component Analysis* (Springer, Berlin, 1986)
2. M.A. Turk, A.P. Pentland, Face recognition using eigenfaces, in *IEEE International Conference on Computer Vision and Pattern Recognition CVPR (1991)*
3. I. Borg, P. Groenen, *Modern Multidimensional Scaling: Theory and Applications* (Springer, Berlin, 2005)
4. P. Belhumeur, J. Hespanha, D. Kriegman, Eigenface vs. fisherfaces:recognition using class specific linear projection. IEEE Trans. Pattern Anal. Mach. Intell. **19**, 711–720 (1997)

5. J.B. Tenenbaum, V. De Silva, J.C. Langford, A global geometric framework for nonlinear dimensionality reduction. Science **290**(5500), 2319–2323 (2000)
6. S. Lafon, A.B. Lee, Diffusion maps and coarse-graining: a unified framework for dimensionality reduction, graph partitioning, and data set parameterization. IEEE Trans. Pattern Anal. Mach. Intell. **28**(9), 1393–1403 (2006)
7. M. Belkin, P. Niyogi, Laplacian eigenmaps and spectral techniques for embedding and clustering, in *Advances in Neural Information Processing System*, vol. 16 (MIT, 2004), p. 153
8. S.T. Roweis, L.K. Saul, Nonlinear dimensionality reduction by locally linear embedding. Science **290**, 2323–2326 (2000)
9. D.L. Donoho, C. Grimes, Hessian eigenmaps: locally linear embedding techniques for high-dimensional data. Proc. Natl. Acad. Sci. 100(10), 5591–5596 (2003)
10. Z. Zhang, H. Zha, Principal manifolds and nonlinear dimensionality reduction by local tangent space alignment. SIAM J. Sci. Comput. **26**, 313–338 (2004)
11. H. Li, T. Jiang, K. Zhang, Efficient and robust feature extraction by maximum margin criterion. IEEE Trans. Neural Networks 17(1), 157–165 (2006)
12. F. Wang, C. Zhang, Feature extraction by maximizing the average neighborhood margin. in *IEEE Conference on Computer Vision and Pattern Recognition, 2007* (CVPR'07) (IEEE, 2007), pp. 1–8
13. D. Zhao, Z. Lin, X. Tang Classification via semi-Riemannian spaces. in IEEE Conference on Computer Vision and Pattern Recognition, 2008 (CVPR 2008) (IEEE, 2008), pp. 1–8
14. X. Niyogi, Locality preserving projections, in *Neural Information Processing Systems*, vol. 16 (MIT, 2004), p. 153
15. X. He, S. Yan, Y. Hu, P. Niyogi, H.-J. Zhang, Face recognition using Laplacianfaces. IEEE Trans. Pattern Anal. Mach. Intell. **27**(3), 328–340 (2005)
16. J. Yang, D. Zhang, J.-Y. Yang, B. Niu, Globally maximizing, locally minimizing: unsupervised discriminant projection with applications to face and palm biometrics. IEEE Trans. Pattern Anal. Mach. Intell. **29**(4), 650–664 (2007)
17. X. He, D. Cai, S. Yan, H.-J. Zhang, Neighborhood preserving embedding. in *Tenth IEEE International Conference on Computer Vision, 2005* (ICCV 2005), vol. 2 (IEEE, 2005), pp. 1208–1213
18. D. Zhao, Z. Lin, X. Tang, Laplacian pca and its applications, in *IEEE 11th International Conference on Computer Vision, 2007 (ICCV 2007)* (IEEE, 2007), pp. 1–8
19. S. Yan, D. Xu, B. Zhang, H.J. Zhang, Q. Yang, S. Lin, Graph embedding and extensions: a general framework for dimensionality reduction. IEEE Trans. Pattern Anal. Machine Intell. **27**, 40–51 (2007)
20. H.-T. Chen, H.-W. Chang, T.L. Liu, Local discriminant embedding and its variants, in *IEEE Computer Society Conference on Computer Vision and Pattern Recognition, 2005 (CVPR 2005)*, vol. 2 (IEEE, 2005), pp. 846–853
21. J.M. Lee, *Riemannian Manifolds: An Introduction to Curvature* (Springer, Berlin, 1991)
22. B. O'Neill, *Semi-Riemannian Geometry with Applications to Relativity* (Academic Press, New York, 1983)
23. P.J. Phillips, P.J. Flynn, T. Scruggs, K.W. Bowyer, J. Chang, K. Hoffman, J. Marques, J. Min, W. Worek, Overview of the face recognition grand challenge, in *IEEE Computer Society Conference on Computer Vision and Pattern Recognition, 2005 (CVPR 2005)*, vol. 1 (IEEE, 2005), pp. 947–954

Chapter 9
Curve Interpolation and Financial Curve Construction

Pengfei Huang, Haiyan Wang, Ping Wu, and Yifei Li

Abstract In this chapter, first we introduce some commonly used curve interpolation methods for interest rate curves. Then, we present a positivity-preserving piecewise rational cubic interpolation function. It is constructed to ensure positive values by adjusting the shape-control parameters. When it is applied to the interest rate curve construction, this interpolation algorithm ensures positive values. The market data has been reconstructed to restrict the fluctuation of the interest rate curve when the market data changes sharply.

9.1 Introduction

Interest rate curves are the foundation of the pricing of interest rate derivatives, and reflect the market's expectation of future interest rate development [1]. The market interest rate data is usually a set of discrete time sequences with different gaps, by which one can construct an interest rate curve with various interpolation algorithms. These algorithms should guarantee the interest rate curve satisfies continuity and smoothness conditions and accurately reflects the current market.

The cubic spline interpolation method [2] proposed by McCulloch (1975) is a popular choice for interest rate curve construction, and was developed by Fisher et al. (1995) to ensure the smoothness and goodness of fit of the interest rate curve [3]. However, the interest rate curve constructed by cubic spline interpolation is not guaranteed to be positive and is not arbitrage free. Lu Jun and Han Xuli (2005) proposed a rational cubic spline with cubic denominator and numerator which can preserve the monotonicity of the inputs [4]. Tian Meng (2006) proposed a rational cubic spline with cubic numerator and quadratic denominator which can guarantee the positivity of the constructed curve [5]. Based on the existence of rational splines,

P. Huang • H. Wang (✉)
School of Economics and Management, Southeast University, Nanjing, China
e-mail: 220141843@seu.edu.cn; hywang@seu.edu.cn

P. Wu • Y. Li
BMO Financial Group, Bank of Montreal, Toronto, ON, Canada
e-mail: Ping.Wu@bmo.com; Stephanie.li@bmo.com

© Springer International Publishing Switzerland 2015
L.M. Chen et al., *Mathematical Problems in Data Science*,
DOI 10.1007/978-3-319-25127-1_9

in this paper we propose a rational cubic spline with a cubic denominator and numerator that can ensure the constructed interest rate curve is positive.

9.2 Basic Interpolation Methods for Interest Rate Curves

The term structure of interest rates is fundamental to consistently value interest rate instruments. The term structure can refer to different curves: the discount curve, zero curve, or forward curve. However, we cannot obtain these curves directly from market data; instead, we use discrete dates and market quotes, and a wide range of models and numerical methods are employed to fit a continuous function to a set of discretely observed quotes.

As a brief review, we provide the mathematical background and notation for the term structure. There are three equivalent descriptions: the discount function $d(t)$, the yield curve $r(t)$, and instantaneous forward rate curve $f(t)$.

$$d(t) = e^{-r(t)t}$$

is equivalent to

$$d(t) = e^{-\int_0^t f(u)du}$$

where

$$f(t) = \frac{\partial}{\partial t} r(t) t.$$

9.2.1 Linear Interpolation

Suppose we have market quotes $r_1, r_2, \ldots, r_n$ at times $t_1, t_2, \ldots, t_n$. The objective is to determine r_t for any time $t \in [t_1, t_n]$. A straightforward method is linear interpolation. The interpolation can be linear on rates, discount factors, and the transformation of discount factors. Some common approaches are as follows:

First, we can give a formula for the zero rates. We determine i for time t such that $t \in [t_i, t_{i+1}]$, and r_t is calculated as the linear interpolation of r_i and r_{i+1}:

$$r_t = \frac{(t_{i+1} - t)}{(t_{i+1} - t_i)} r_i + \frac{(t - t_i)}{(t_{i+1} - t_i)} r_{i+1}.$$

For the instantaneous forward curve $f(t) = \frac{\partial}{\partial t} r(t) t$, we have

$$f(t) = \frac{(t_{i+1} - 2t)}{(t_{i+1} - t_i)} r_i + \frac{(2t - t_i)}{(t_{i+1} - t_i)} r_{i+1}.$$

When it is linear on the discount factors, we can apply the linear interpolation on the discount factors directly. Since the discount factors are defined as $d(t) = e^{-r(t)t}$, we have a set of discount factors $d_1, d_2, \ldots, d_n$ corresponding to $r_1, r_2, \ldots, r_n$. The formula is

$$d(t) = \frac{(t_{i+1} - t)}{(t_{i+1} - t_i)} d_i + \frac{(t - t_i)}{(t_{i+1} - t_i)} d_{i+1}.$$

According to

$$r(t) = \frac{-\ln(d(t))}{t}$$

we have

$$r(t) = \frac{-1}{t} \ln \left(\frac{(t_{i+1} - t)}{(t_{i+1} - t_i)} d_i + \frac{(t - t_i)}{(t_{i+1} - t_i)} d_{i+1} \right).$$

Then, the forward rate can be calculated as

$$f(t) = \frac{(d_i - d_{i+1})}{((t_{i+1} - t)d_i + (t - t_i)d_{i+1})}. \tag{9.1}$$

If discount factors are linear on the logarithmic rate, we can derive a formula accordingly.

9.2.2 Cubic Spline Interpolation

As well as considering the same set of assumptions as in linear interpolation, spline interpolation requires a separate cubic polynomial for each interval. The aim is to determine the coefficients and the value of the rate at time t:

$$r(t) = a_i + b_i(t - t_i) + c_i(t - t_i)^2 + d_i(t - t_i)^3, \quad t \in [t_i, t_{i+1}].$$

In order to determine the set of coefficients $\{a_i, b_i, c_i, d_i, \ i = 1, 2, \ldots, n - 1\}$, constraints on the value, the first derivative, and the second derivatives at t_i should be satisfied.

First, the value of the piecewise polynomial should be identical at the knots (t_i), which gives the following system of equations:

$$r_i = a_i,$$

$$r_{i+1} = a_i + b_i(t_{i+1} - t_i) + c_i(t_{i+1} - t_i)^2 + d_i(t_{i+1} - t_i)^3.$$

Since

$$f(t) = \frac{\partial}{\partial t} r(t)t = \frac{\partial r(t)}{\partial t} + r(t),$$

and

$$\frac{\partial r(t)}{\partial t} = b_i + 2c_i(t - t_i) + 3d_i(t - t_i)^2, \quad t \in [t_i, t_{i+1}],$$

the piecewise polynomial should be differentiable to guarantee the continuity of the instantaneous forward rate curve $f(t)$. Thus, we have

$$b_{i+1} = b_i + 2c_i(t_{i+1} - t_i) + 3d_i(t_{i+1} - t_i)^2, \quad i = 1, 2, \ldots, n - 2.$$

Combining the $2(n-2)$ conditions above, we now have $3n - 4$ equations for $4n - 4$ parameters. We still need another n constraints to obtain all of the parameters. We can use the so-called natural cubic splines to do this based on the second derivatives:

$$\frac{\partial^2 r(t)}{\partial t^2} = 2c_i + 6d_i(t - t_i), \quad t \in [t_i, t_{i+1}].$$

These cubic splines are chosen so that the second derivatives of the functions at the knots are continuous and the second derivatives of the end points equal zero, which gives n linear equations in the system:

$$c_{i+1} = c_i + 3d_i(t_{i+1} - t_i), \quad i = 1, 2, \ldots, n - 2,$$

$$c_1 = 0,$$

$$c_{n-1} + 3d_{n-1}(t_n - t_{n-1}) = 0.$$

These splines guarantee the continuity and smoothness of the curve.

9.2.3 Monotone and Convex Splines

A simple observation is that the cubic splines in the last subsection could unexpectedly generate negative forward rates. The monotone convex spline method is introduced to build a continuous, smooth curve that ensures positive forward rates between the observations. Thus, the curve is arbitrage free. First, we calculate the forward rates $f_i^d, i = 1, 2, \ldots, n$ from the market quotes $r_1, r_2, \ldots, r_n$ at times $t_1, t_2, \ldots, t_n$, assuming forward rates are constant in each interval and are continuously compounded [6].

Various methods can be applied to fit a curve to the term structure of the interest rate; goodness of fit and smoothness of the curves are two measures of success. All such methods have strengths and weaknesses, and their selection depends on the specific requirement of the application.

In the following, we propose a rational cubic interpolation method for positive forward rates, which we call positivity preservation.

9.3 Rational Cubic Interpolation Method for Positivity Preservation

Let $a = \tau_1 < \tau_2 < \ldots < \tau_n = b$ be a set of time knots, $\theta = \frac{\tau - \tau_i}{\tau_{i+1} - \tau_i}$, $h_i = \tau_{i+1} - \tau_i$, and $f(\tau_i) = f_i (i = 1, 2, \ldots, n)$ be the input data at time knot τ_i. Define the rational cubic spline interpolation function in the ith interval $[\tau_i, \tau_{i+1}]$ as

$$S_i(\tau) = \frac{P_i(\theta)}{Q_i(\theta)}, \quad i = 1, 2, \ldots, n-1, \tag{9.2}$$

where

$$P_i(\theta) = (1 - \theta)^3 f_i + \theta(1 - \theta)^2 A + \theta^2 (1 - \theta) B + \theta^3 f_{i+1} \tag{9.3}$$

$$Q_i(\theta) = (1 - \theta)^3 + v_i(1 - \theta)^2 \theta + u_i(1 - \theta)\theta^2 + \theta^3. \tag{9.4}$$

In Eq. (9.4) $v_i > 0$ and $u_i > 0$ are shape-control parameters, and d_i is the first derivative of the function $f(\tau)$ at time knot τ. To fulfill the condition $S(x) \in C^1[a, b]$, the interpolation function defined by Eq. (9.2) should satisfy the following conditions:

$$\begin{cases} S_i(\tau_i) &= f_i \\ S_i(\tau_{i+1}) &= f_{i+1} \\ S_i'(\tau_i) &= d_i \\ S_i'(\tau_{i+1}) &= d_{i+1}. \end{cases} \tag{9.5}$$

The terms A and B are given by

$$\begin{cases} A = d_i h_i + v_i f_i \\ B = -d_{i+1} h_i + u_i f_{i+1}. \end{cases} \tag{9.6}$$

The first derivative at time knot τ_i, d_i, is unknown in the market and is estimated as follows. We take two knots from both sides of τ_i (if one side does not have two, we take more from the other side). Let $\Delta \tau_j = \tau_j - \tau_i$, assume the input data at the knots is $f_{(j)} (j = 1, 2, 3, 4)$, and define

$$\begin{cases} \lambda_0 = -\sum_{j=1}^{4} \lambda_j \\ \lambda_j = \dfrac{\prod_{k=1}^{4} \Delta\tau_k}{\Delta\tau_j^2 \prod_{\substack{k=1 \\ k \neq j}}^{4} (\Delta\tau_k - \Delta\tau_j)}, \quad j = 1,2,3,4. \end{cases} \tag{9.7}$$

Then, d_i can be estimated by

$$d_i = \lambda_0 f_i + \sum_{j=1}^{4} \lambda_j f_{(j)}. \tag{9.8}$$

The accuracy of the estimation is $O(\max_{1 \leq j \leq 4} |\Delta\tau_j^4|)$ [7], which is good. This will improve the smoothness and help retain the monotonicity of the inputs to the interest rate curve. However, as seen in Eq. (9.8), the estimation of the first derivative at knot τ_i is achieved using four nearby points, so changes at knot τ_i affect the shape of the curve considerably. The positivity-preserving property is analyzed as follows. Since $\theta = \frac{\tau - \tau_i}{\tau_{i+1} - \tau_i}$ and $\tau \in [\tau_i, \tau_{i+1}]$, θ is in the interval $[0, 1]$. Hence, a sufficient condition for the positivity-preserving property of the interpolation function defined by Eq. (9.2) is $A > 0, B > 0$, and the value range of the shape-control parameters can be obtained as follows:

$$\begin{cases} v_i > \max(-\frac{d_i h_i}{f_i}, 0), \\ u_i > \max(\frac{d_{i+1} h_i}{f_{i+1}}, 0). \end{cases} \tag{9.9}$$

When v_i and u_i satisfy these conditions (9.9), the curve constructed with the rational cubic spline interpolation method will always be positive and arbitrage free. The error estimation of the rational cubic spline interpolation is analyzed as follows. It is sufficient to only deal with the case where the knots are equally spaced. Without loss of generality, we only consider the subintervals $[\tau_i, \tau_{i+1}]$ $(3 \leq i \leq n - 2)$, and when the knots are equally spaced let $h_i = h$. By Eqs. (9.7) and (9.8), we get

$$\lambda_1 = \frac{1}{12h}, \lambda_2 = -\frac{2}{3h}, \lambda_3 = \frac{2}{3h}, \lambda_4 = -\frac{1}{12h}, \lambda_0 = 0, \tag{9.10}$$

$$d_i = \frac{1}{12h} f_{i-2} - \frac{2}{3h} f_{i-1} + \frac{2}{3h} f_{i+1} - \frac{1}{12h} f_{i+2}. \tag{9.11}$$

Then, the numerator of the interpolation function in Eq. (9.3) can be transformed into the following form:

$$P_i(\theta) = l_1 f_{i-2} + l_2 f_{i-1} + l_3 f_i + l_4 f_{i+1} + l_5 f_{i+2} + l_6 f_{i+3}, \tag{9.12}$$

where

$$\begin{cases} l_1 = \frac{1}{12}\theta(1-\theta)^2 \\ l_2 = \theta(1-\theta)[-\frac{2}{3}(1-\theta) - \frac{1}{12}\theta] \\ l_3 = (1-\theta)[(1-\theta)^2 + \theta(1-\theta)v_i + \frac{2}{3}\theta^2] \\ l_4 = \theta[\theta(1-\theta)u_i + \theta^2 + \frac{2}{3}(1-\theta)^2] \\ l_5 = -\theta(1-\theta)[\frac{1}{12}(1-\theta) + \frac{2}{3}\theta] \\ l_6 = \frac{1}{12}\theta^2(1-\theta). \end{cases} \tag{9.13}$$

When $f(t) \in C^1[a, b]$ and $S(t)$ is the rational cubic interpolation function of $f(t)$ in the subinterval $[\tau_i, \tau_{i+1}]$, using the Peano–Kernel Theorem gives

$$R[f] = f(t) - S(t) = \int_{\tau_{i-2}}^{\tau_{i+3}} f'(\tau) R_t[(t - \tau)_+^0] d\tau, \qquad (9.14)$$

where

$$R_t[(t - \tau)_+^0] = \begin{cases} 1 - \frac{l_2 + l_3 + l_4 + l_5 + l_6}{Q_i(\theta)}, & \tau_{i-2} < \tau < \tau_{i-1} \\[2mm] 1 - \frac{l_3 + l_4 + l_5 + l_6}{Q_i(\theta)}, & \tau_{i-1} < \tau < \tau_i \\[2mm] 1 - \frac{l_4 + l_5 + l_6}{Q_i(\theta)}, & \tau_i < \tau < t \\[2mm] -\frac{l_4 + l_5 + l_6}{Q_i(\theta)}, & t < \tau < \tau_{i+1} \\[2mm] -\frac{l_5 + l_6}{Q_i(\theta)}, & \tau_{i+1} < \tau < \tau_{i+2} \\[2mm] -\frac{l_6}{Q_i(\theta)}, & \tau_{i+2} < \tau < \tau_{i+3}. \end{cases} \qquad (9.15)$$

From Eq. (9.14) we get the following result:

$$|R[f]| = |\int_{\tau_{i-2}}^{\tau_{i+3}} f'(t) R_t[(t - \tau)_+^0] d\tau| \le |f'(t)| [\int_{\tau_{i-2}}^{\tau_{i-1}} |1 - \frac{l_2 + l_3 + l_4 + l_5 + l_6}{Q_i(\theta)}| d\tau$$

$$+ \int_{\tau_{i-1}}^{\tau_i} |1 - \frac{l_3 + l_4 + l_5 + l_6}{Q_i(\theta)}| d\tau + \int_{\tau_i}^{t} |1 - \frac{l_4 + l_5 + l_6}{Q_i(\theta)}| d\tau$$

$$+ \int_{t}^{\tau_{i+1}} |-\frac{l_4 + l_5 + l_6}{Q_i(\theta)}| d\tau + \int_{\tau_{i+1}}^{\tau_{i+2}} |-\frac{l_5 + l_6}{Q_i(\theta)}| d\tau$$

$$+ \int_{\tau_{i+2}}^{\tau_{i+3}} |-\frac{l_6}{Q_i(\theta)}| d\tau] \le h|f'(t)| w(v_i, u_i, \theta), \quad (9.16)$$

where

$$w(v_i, u_i, \theta) = 3 + \frac{1}{Q_i(\theta)}(|l_2 + l_3 + l_4 + l_5 + l_6| + |l_3 + l_4 + l_5 + l_6| + 2|l_4 + l_5 + l_6| + |l_5 + l_6| + |l_6|).$$
$$(9.17)$$

For a given v_i and u_i, let $e_i = \max_{0 < \theta < 1} w(v_i, u_i, \theta)$. This term is independent of the subinterval $[\tau_i, \tau_{i+1}]$, and only dependent on the parameters v_i and u_i. The values

Table 9.1 The values of e_i for $v_i = 0.001$ with various values of u_i

v_i	u_i	e	θ
0.001	0.001	6.9862	0.7040
0.001	0.005	6.9862	0.7040
0.001	0.010	6.9863	0.7040
0.001	0.050	6.9865	0.7039
0.001	0.100	6.9867	0.7037
0.001	1.000	6.9901	0.7015
0.001	10.000	6.9972	0.6974
0.001	100.000	6.9997	0.6961
0.001	1000.000	7.0000	0.6959

Table 9.2 The values of e_i for $u_i = 0.001$ with various values of v_i

v_i	u_i	e	θ
0.005	0.001	6.9849	0.7060
0.010	0.001	6.9833	0.7086
0.050	0.001	6.9720	0.7322
0.100	0.001	7.0000	0.9999
1.000	0.001	7.0000	1.0000
10.000	0.001	7.0000	0.9999
100.000	0.001	7.0000	0.9999
1000.000	0.001	7.000	0.9999

e_i, for various v_i and u_i, are given in Tables 9.1 and 9.2. Based on the experimental results in Tables 9.1 and 9.2, we can see the value of e_i only changes slightly for $v_i > 0$ and $u_i > 0$. This demonstrates that this interpolation method is stable.

9.4 Reconstruction of the Market Interest Rate Data

Table 9.3 details the USD market data from 04/01/2000. In Table 9.3, the first column is the time from start to expiration, the second column is the expiration date, the third column is the market rate, and the last column is the time in years (365 days=1 year). To restrict the fluctuation of the interest rate curve when the input data changes sharply, we reconstruct the market data before applying the rational cubic interpolation method. Let (τ_i, r_i) $(i = 1, 2, \ldots, n)$ be a given set of market data, where τ_i are the time knots. We assume r_i is the value corresponding to the entire subinterval $[\tau_{i_1}, \tau_i]$ $(i = 2, \ldots, n)$. The reconstructed rate is f_i at the point τ_i $(i = 2, \ldots, n - 1)$:

$$f_i = \frac{\tau_i - \tau_{i-1}}{\tau_{i+1} - \tau_{i-1}} r_{i+1} + \frac{\tau_{i+1} - \tau_i}{\tau_{i+1} - \tau_{i-1}} r_i. \tag{9.18}$$

Table 9.3 USD market data in 2000

Time length	Expiration date	Market interest r (%)	Time length τ
7 day	13-Jan-00	5.53125	0.019
1 month	7-Feb-00	5.81250	0.088
3 months	6-Apr-00	6.03125	0.249
6 months	6-Jul-00	6.21875	0.499
12 months	8-Jan-01	6.59375	1.008
2 yr	7-Jan-02	6.8950	2.005
3 yr	6-Jan-03	7.0250	3.003
4 yr	6-Jan-04	7.0850	4.003
5 yr	6-Jan-05	7.1350	5.005
6 yr	6-Jan-06	7.1750	6.005
7 yr	8-Jan-07	7.2250	7.011
8 yr	7-Jan-08	7.2650	8.008
9 yr	6-Jan-09	7.2950	9.008
10 yr	6-Jan-10	7.3350	10.008
12 yr	6-Jan-12	7.3850	12.008
15 yr	6-Jan-15	7.4350	15.011
20 yr	6-Jan-20	7.4450	20.014
25 yr	6-Jan-25	7.4450	25.019
30 yr	7-Jan-30	7.4350	30.025

We add the additional subintervals $[\tau_0, \tau_1]$ and $[\tau_n, \tau_{n+1}]$ at the start and end, where

$$\begin{cases} \tau_0 &= \tau_1 - (\tau_2 - \tau_1) \\ \tau_{n+1} &= \tau_n + (\tau_n - \tau_{n-1}) \end{cases} \tag{9.19}$$

and

$$r_{n+1} = r_n + \frac{\tau_n - \tau_{n-1}}{\tau_n - \tau_{n-2}}(r_n - r_{n-1}). \tag{9.20}$$

Then, the reconstructed market data f_i ($i = 1, 2, \ldots, n$) can be obtained from Eq. (9.18) [6]. The results of applying the reconstruction algorithm to the market data (from Table 9.3) are given in Table 9.4. The third column in Table 9.4 is the reconstructed market data.

9.5   Analysis of the Interest Curve Constructed by Rational Cubic Interpolation

The constructed interest rate curve, found by applying the rational cubic interpolation method to the reconstructed market data, is given in Fig. 9.1. In general, the term structure contains more short-term information, so the short term of the

Table 9.4 USD market data
in 2000

Serial number	Time length τ	Market rare f
1	0.019	0.05671875
2	0.088	0.05878125
3	0.249	0.06104699
4	0.499	0.06342268
5	1.008	0.06695567
6	2.005	0.06959967
7	3.003	0.07054970
8	4.003	0.07109975
9	5.005	0.07155020
10	6.005	0.07199925
11	7.011	0.07245090
12	8.008	0.07279977
13	9.008	0.07315000
14	10.008	0.07351667
15	12.008	0.07404988
16	15.011	0.07438751
17	20.014	0.07445000
18	25.019	0.07440000
19	30.025	0.07432500

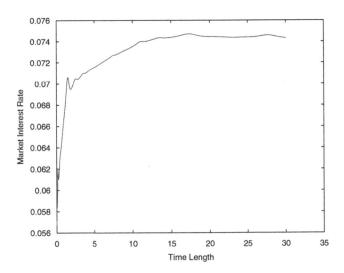

Fig. 9.1 Interest curve constructed by applying rational cubic interpolation to the reconstructed market interest rate data

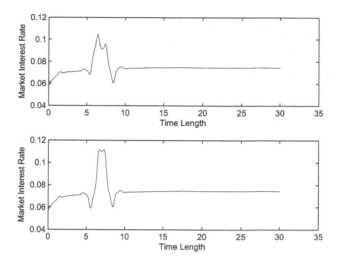

Fig. 9.2 The interest curve constructed by rational cubic interpolation with and without the reconstruction of market interest rate data

interest rate curve should focus on goodness of fit while the long term of the curve requires smoothness [8]. The interest rate curve constructed by the rational cubic interpolation method essentially satisfies these requirements. We now discuss the stability of the interest rate curve; when rational cubic interpolation is applied to the reconstructed market data the data in Table 9.3 changes from $(7.011, 7.2250\%)$ to $(7.011, 11.0000\%)$. The two interest rate curves in Fig. 9.2 are constructed by the rational cubic interpolation method; the upper curve uses reconstructed market data and the lower curve uses the original market data. It can be seen that the upper curve, from reconstructed market data, has less volatility.

9.6 Conclusion

The rational cubic interpolation algorithm can ensure that the constructed interest rate curve is positivity-preserving. The interpolation function is C^1 continuous and the constructed interest rate curve is stable and smooth. The algorithm to reconstruct the data can restrict the fluctuations of the interest rate curve when the input data jumps sharply.

References

1. Y. Gu, Z. Zhu, Y. Yan, The term structure and macro-economy of China. China Financ. **3**, 87 (2014)
2. J.H. McCulloch, The tax-adjusted yield curve. J. Financ. **30**, 811–830 (1975)

3. M. Fisher, D. Nuchka, D. Zervos, Fitting the term structure of interest rates with smoothing splines. Comput. Graph. **15**, 14–23 (1995)
4. L. Jun, H. Xuli, Monotonicity preserving rational cubic spline interpolation. Math. Theory Appl. **25**(3), 115–117 (2005)
5. T. Meng, Positivity-preserving piecewise rational cubic interpolation. J. Shandong Univ. Technol. **20**(3), 16–18 (2006)
6. P.S. Hagan, G. West, Interpolation methods for curve construction. Appl. Math. Financ. **13**(2), 89–129 (2006)
7. N. Wang, Piecewise spline interpolation. J. Huazhong Coll. Technol. **3**, 50–59 (1975)
8. H. Hu, Z. Fang, The improvement of the smoothing spline of the term structure of China. J. Manage. Sci. China **12**(1), 101–111 (2009)

Chapter 10
Advanced Methods in Variational Learning: Segmentation with Intensity Inhomogeneity

Jack Spencer and Ke Chen

Abstract Mathematical imaging aims to develop new mathematical tools for the emerging field of image and data sciences. Automatic segmentation in the variational framework is a challenging as well as a demanding task for a number of imaging tasks. Achieving robustness and reliability is a major problem. The two-phase piecewise-constant model of the Mumford–Shah (2PCMS) energy is most suitable for images with simple and homogeneous features where the intensity variation is limited. However, it has been applied to many different types of synthetic and real images after some adjustments to the formulation. This chapter addresses the task of generalising this widely used model to account for intensity inhomogeneity which is common for real life images. We consider two ways of altering the fitting terms by 'correcting' the intensity. This allows us to process the type of images that are beyond 2PCMS. We first review the state of the art of such methods for treating inhomogeneity and then demonstrate inconsistencies in existing methods. Next we propose two modified variational models by introducing additional constraints and extend the concept to selective segmentation, where user input can aid the intensity correction. These models are minimised with convex relaxation methods, where the global minimiser can be found for a fixed fitting term. Finally, we present numerical results that demonstrate an improvement to existing methods in terms of reliability.

J. Spencer • K. Chen (✉)
Centre for Mathematical Imaging Techniques and Department
of Mathematical Sciences, University of Liverpool, Merseyside, UK
e-mail: j.a.spencer@liv.ac.uk; k.chen@liv.ac.uk
http://www.liv.ac.uk/~cmchenke

© Springer International Publishing Switzerland 2015   171
L.M. Chen et al., *Mathematical Problems in Data Science*,
DOI 10.1007/978-3-319-25127-1_10

10.1   Introduction

Segmentation is a fundamental aspect of image processing, and variational methods
have proved to be a very effective approach to this important problem. In Sect. 10.2
we introduce the concept of image segmentation, its major applications, and the
areas of interest to this work. In Sect. 10.3 we provide a background for variational
methods for image segmentation, by detailing two foundational formulations:
Mumford–Shah [17], and Chan–Vese [6]. We set out the focus and motivation of our
work, before detailing the Variant Mumford–Shah model (VMS) [8] in Sect. 10.4.
We discuss the problems with recovering the 'true' image, in particular the lack
of convergence of its constituent parts. In Sect. 10.5 we detail the introduction
of constraints to VMS in order to ensure the convergence of all variables being
minimised, as well as ensuring their feasibility in relation to the image. We have two
approaches: restricting the size of the intensity constants by an initial approximation
(Method 1), and restricting the scale of the bias field (Method 2). In Sect. 10.6
we include experimental results that measure the accuracy of the new methods
compared to VMS by using the Tannimoto Coefficient [9], and discuss the benefits
of these approaches in Sect. 10.7.

10.2   Image Segmentation

The task of partitioning an image into multiple regions each sharing certain
characteristics—such as texture, intensity, shape or colour—is called segmentation,
and is an important aspect of image processing. Given an image $z(x, y)$ in a bounded
domain $\Omega \subset \mathbb{R}^2$, we look for an edge Γ that partitions Ω into regions $\{\Omega_i, i =
1, 2, \ldots, N\}$ in $\Omega \setminus \Gamma$. Typically, the number of regions is understood to be known a
priori, which significantly simplifies the problem. In this work we consider the case
$N = 2$, such that we are determining foreground and background regions.

There are many applications of segmentation in a broad range of fields. Medical
imaging is an important aspect of medical research and patient care, being a
major part of clinical diagnosis. Segmentation is an essential part of this field,
allowing for an improved analysis of various image types such as PET (positron
emission tomography), MRI (magnetic resonance imaging) and CT. Currently,
the interpretation of these images has a critical impact on patient care. The
incorporation of image segmentation into the diagnosis and treatment process allows
for improved outcomes due to quantification and precision aiding the subjective
interpretation of healthcare professionals in a number of ways. Segmentation also
aids aspects of the security sector. Event recognition and object tracking can help
in many different areas such as securing private property or sensitive locations.
Automating this process has substantial benefits to these industries as it provides
an efficient alternative to traditional methods. Image segmentation is also central
to the robotics field, helping automatic visual sensing, fundamental to autonomous
navigation.

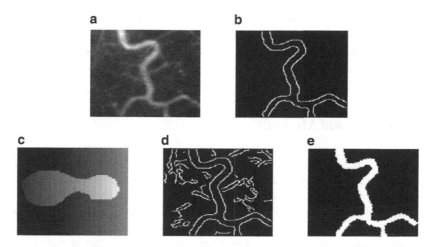

Fig. 10.1 The *top row* is a piecewise-constant image, with a successful result using the 'edge' function. The *second row* is for an image with background artefacts and slight noise. The 'edge' function fails, whereas the 'activecontour' function succeeds. Both functions are included in MATLAB's Image Processing Toolbox. (**a**) Piecewise-constant image. (**b**) Edge detection. (**c**) Image. (**d**) Failed edge detection. (**e**) 'Activecontour' result

There are many different approaches to tackle the problem of partitioning an image into relevant regions. The aspect of interest here is to primarily consider intensity as a segmentation criterion. For simple examples, where there is little variation of intensity within the foreground and background, it is possible to simply threshold the image to determine regions of interest, Ω_1 and Ω_2. Another option is to use edge detectors to determine the correct partition Γ, such as the 'edge' function in MATLAB's Image Processing Toolbox. Figure 10.1a–b shows such an example, where there is a piecewise-constant image, and the resulting edge function clearly gives the correct partition Γ. However, simplistic approaches like edge detection or thresholding have limited use with real images, that also include noise, blur, inhomogeneity and artefacts. In these cases, the boundaries computed are often incomplete such that they do not completely define objects, or pick up more detail than required.

There are many alternatives to tackling more difficult segmentation problems, rather than the naive approaches mentioned above and demonstrated in Fig. 10.1e. One such option is the region growing approach, where image statistics determine the splitting of the image. However, the boundaries computed can often be irregular and can be sensitive to initialisation. There are also combinatorial methods, that view segmentation as a discrete label assignment problem, but there can be limitations in terms of the type of data terms that can be accommodated. We consider the variational approach, which we introduce next. In this way, an energy functional that describes the image data is minimised, and the solution gives the correct boundary Γ, and corresponding foreground and background regions. An important example of this type of model is Active Contours Without Edges [6] by Chan and

Vese, which we detail in Sect. 10.3. Figure 10.1 demonstrates a successful result using the 'activecontour' function in MATLAB's Image Processing Toolbox, where edge detection fails. The image contains noise and background artefacts, which can be treated by region based variational segmentation models easily.

10.3 Variational Methods

Variational image segmentation models are defined in a continuous setting and formulated as energy minimisation problems. The approach is to define an energy that describes some characteristics of the image and then find an optimal function, that defines the unknown edge or regions of interest. Two main approaches to segmentation by image intensity have developed. One is based on edges, where edge detectors are incorporated into an energy functional, and the other is region based, where the intensity throughout the image is considered. We first briefly review the development of edge based active contour models, and then discuss the focus of this work: region based methods.

One of the most important models in variational image segmentation, proposed by Kass et al. [12] in 1988, minimises an objective functional towards a regular curve on edges, and is known as Snakes [12]. The energy characterises the contour and controls the elastic and bending energies of a piecewise $\mathscr{C}^1$ planar curve. An additional term, known as the external energy, employs an edge function to attract the contour towards edges within the image, z. When employed naively edge detectors rarely produce successful results, as shown in Fig. 10.1. However, when embedded into the variational formulation they can prove very effective. The local minima of this functional can be found by computing the corresponding Euler–Lagrange equation. Caselles et al. [4] proposed the geodesic active contour model in 1997, a particular case of the Snakes model. The functional only involves one term, again reliant on an edge function to locate boundaries. These active contour models [4, 12] depend on the gradient of the image, and have been widely studied since their inception proving very successful. However, when applied to some real images the edge function is not zero at the edges and the curve can pass through object boundaries. This can be counteracted by smoothing the image, which can in turn lead to the loss of edge definition and limited results. These problems can be avoided with an alternative approach; region based methods.

Region based methods tend to be more flexible in terms of their application, and more robust to phenomena that occur in real images: noise, blur, inhomogeneity and artefacts. These approaches consider image intensities throughout the image, rather than focus on sharp intensity jumps. The focus of this work, and a fundamental idea in variational image segmentation, is the introduction of the minimisation of the Mumford–Shah functional [17], given by

$$E^{MS}(u, \Gamma) = \lambda \int_{\Omega} (z - u)^2 \, d\Omega + \int_{\Omega \backslash \Gamma} |\nabla u| \, d\Omega + \nu |\Gamma|, \qquad (10.1)$$

where $\lambda, \nu > 0$ are weighting parameters, and $|\Gamma|$ denotes the length of the edge curve Γ, the boundary between regions Ω_i. Mumford and Shah [17] demonstrated that, theoretically, the existence and regularity of minimisers of (10.1) can be achieved. The function that minimises (10.1) is a piecewise smooth function we denote u_{MS}, that has discontinuities given by Γ, defining object boundaries. In 2001, Chan and Vese introduced Active Contours Without Edges [6] and a functional that was a particular case of (10.1): the two-phase piecewise-constant example, i.e. $|\nabla u| = 0$ and $N = 2$ in (10.1):

$$E^{CV}(c_1, c_2, \Gamma) = \lambda \left(\int_{\Omega_1} (z - c_1)^2 \, d\Omega + \int_{\Omega_2} (z - c_2)^2 \, d\Omega \right) + \nu |\Gamma|. \tag{10.2}$$

Here, c_1 and c_2 are the average intensities of z in regions Ω_1 and Ω_2, respectively. The Chan–Vese model (CV) has been widely used in segmentation applications since its introduction. It differs from the original Mumford–Shah formulation (10.1) in the sense that it treats the image as piecewise-constant rather than piecewise-smooth. In Fig. 10.2 we include a result for CV, demonstrating that the artefacts in the background of the image are 'ignored' due to the averaging of intensities in each region.

Variational methods involve minimising an energy, such as Eqs. (10.1) and (10.2). In each case the variable that defines the boundary between objects is Γ. To minimise with respect to this variable is not straightforward; topological changes are difficult to track in this framework. An idea that is integral to the development of variational image segmentation models, as well as many other areas in image processing and computer vision, is the level set method, first introduced by Osher and Sethian [19] and Zhao et al. [22]. When applied to segmentation, the level set function $\phi(x, y)$ implicitly represents the one-dimensional contour Γ. By replacing Γ with a two-dimensional function ϕ the motion of this interface ϕ_0 (given by $\phi(x, y) = 0$) is implicit. The geometric characteristics of this interface can be easily tracked with this formulation, as opposed to a parametrisation approach. Caselles et al. [4] minimised their edge based formulation using the variational level set method, Tsai et al. [21] applied it to minimise (10.1), and Chan and Vese

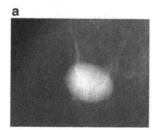

Fig. 10.2 On the *left* is an image containing noise, inhomogeneity, and artefacts. On the *right* is a failed CV result computed with the 'activecontour' function in MATLAB's Image Processing Toolbox. (**a**) Inhomogeneous image. (**b**) Failed 'activecontour' result

applied it to the minimisation of (10.2). Chan and Vese used the Heaviside function to represent the two regions Ω_1 and Ω_2 with respect to the level set function, ϕ, and computed the Euler–Lagrange equation. Despite the considerable success of this model, the solution of the corresponding partial differential equation is often a local minimum, which reduces the reliability of the results. The introduction of the idea of convex relaxation by Chan et al. [7] demonstrated that a global minimum of (10.2) with respect to Γ, and consequently the regions Ω_1 and Ω_2, can be found. The idea is to represent the two regions by an indicator function, and relax the binary constraint such that both the functional and the constraint sets are convex, and has improved the applicability of piecewise-constant segmentation.

We now discuss the main focus of this work; image segmentation with intensity inhomogeneity, based on the CV formulation in particular. Partitioning images by areas of constant intensity (the CV approach) has proved very effective, however can have limited results for more complicated images, as demonstrated in Fig. 10.2. Considering intensity inhomogeneity can offer major benefits and there have been many models that aim to allow for intensity variations within the object, mainly incorporating the level set approach. The CV framework has been generalised by the introduction of new fitting terms to incorporate extensive intensity inhomogeneity, such as Li et al. [14, 15] who introduced a region scalable fitting energy and local cluster method. Jung et al. [11] introduced a nonlocal ACM utilising distance functions. Brox and Cremers [3] and Lanktona and Tannenbaum [13] introduced new local models, incorporating Gaussian kernel functions. Other relevant approaches include Ahmed et al. [1] and Sun et al. [20]. A recent model that combines both the idea of segmentation with intensity inhomogeneity and convex relaxation is D. Chen et al. [8], and our work focuses on their formulation. Their framework is based on the idea of attaining a solution to the piecewise-smooth formulation of Mumford–Shah (10.1), using the piecewise-constant approach of Chan–Vese (10.2). They achieve this with bias field estimation, which we detail next.

10.4   Variant Mumford–Shah Model [8]

D. Chen et al. [8] assume the image can be treated as piecewise-constant. With this assumption, the 'true' image data is formulated [1, 16] as

$$T = \sum_i c_i \chi_i, \quad i = 1, 2, \ldots, N, \tag{10.3}$$

where χ_1 and χ_2 are characteristic functions defining the regions Ω_1 and Ω_2, respectively. The premise is that the image can be modelled as

$$z(x) = B(x)T + \eta, \quad x \in \Omega_i, \quad i = 1, 2, \ldots, N, \tag{10.4}$$

where η is additive noise. Here, as with CV, we consider the two-phase case, i.e. $N = 2$. D. Chen et al. [8] aim to estimate the bias field B and therefore recover the 'true' image T. The VMS [8] is formulated as follows:

$$E^{vms}(c_1, c_2, B) = \lambda \int_{\Omega} \left((z - Bc_1)^2 \chi_1 + (z - Bc_2)^2 \chi_2\right) \, dx + \mu \int_{\Omega} |\nabla B|^2 \, dx + \nu |\Gamma|,$$

(10.5)

where λ, μ and ν are weighting parameters. The idea is that the intensity constants represent the 'true' image, and the bias field B varies such that their combination is the piecewise-smooth approximation of z (with respect to the Mumford–Shah [17] formulation (10.1)), given by

$$u_{MS} = Bc_1 \chi_1 + Bc_2 \chi_2.$$

(10.6)

This is minimised iteratively (e.g. the iterative process method, Li et al. [15]) by the following steps. Step (1): For fixed characteristic functions χ_1 and χ_2, and intensity constants c_1 and c_2, minimise (10.5) with respect to bias field estimator B. Based on the work of Nielsen et al. [18] and Brox and Cremers [3], the exact minimiser can be well approximated. Step (2): For fixed characteristic functions χ_1 and χ_2, and bias field estimator B, minimise (10.5) with respect to intensity constants c_1 and c_2. These can be computed precisely. Step (3): For fixed intensity constants c_1 and c_2, and bias field estimator B, minimise (10.5) with respect to χ_1, χ_2. Minimising two characteristic functions can be achieved by using an indicator function, $u(x)$, which is allowed to take intermediate values. This is based on the work of Chan et al. [7]:

$$\min_{0 \leq u \leq 1} \left\{ \lambda \int_{\Omega} \left((z - Bc_1)^2 - (z - Bc_2)^2\right) u(x) \, dx + \nu \int_{\Omega} |\nabla u(x)| \, dx \right\},$$

(10.7)

where the length term $|\Gamma|$ in (10.5) has been replaced by the total variation of u. D. Chen et al. use the dual formulation method of Chambolle [5], which consists of introducing a new variable v and alternating between minimising u and v. D. Chen at al. [8] implement the three steps given above using slight variations on the formulation.

10.4.1 Convergence Behaviour of VMS

In Fig. 10.3, we demonstrate some results for the VMS that are also used in [8], and are of a comparable quality. However, the question remains: based on the image model described above (10.3), what is the 'true' image? Whilst the joint minimisation of (10.5) with respect to $c_1, c_2, B, \chi_1, \chi_2$ is nonconvex, and therefore we cannot determine the correct c_1 and c_2 precisely, there is a problem with the current framework, which we will now discuss. In Fig. 10.3 (after 300 iterations), we show that the values of the intensity constants continually rise, such that after

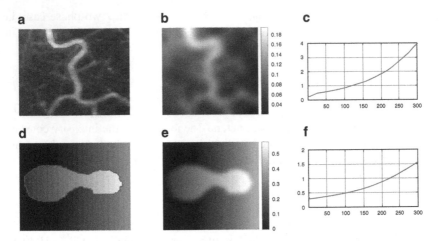

Fig. 10.3 Good results of early termination for VMS (after 300 iterations). From *left to right*: final contour, bias field B, and the progression of c_1 values (vertical axis) against iterations (horizontal axis), demonstrating continuous growth (similar progression for c_2). Note that c_1 and c_2 would indefinitely increase, with B approaching 0. (**a**) Final contour. (**b**) Bias field, B. (**c**) c_1 progression. (**d**) Final contour. (**e**) Bias field, B. (**f**) c_1 progression

2×10^5 iterations, for Image 1, $c_1 = 4.372 \times 10^{66}$ and $c_2 = 3.147 \times 10^{66}$, and for Image 2, $c_1 = 3.422 \times 10^{51}$ and $c_2 = 3.053 \times 10^{51}$. The convergence of u_{MS} (10.6) comes from the reduction in scale of B. To demonstrate this, after the same number of iterations, for Image 1, $||B||_F = 1.023 \times 10^{-65}$, and for Image 2, $||B||_F = 1.513 \times 10^{-50}$ (where $|| \cdot ||_F$ is the Frobenius norm). This motivates our proposal for modifying VMS, in the form of additional constraints, that can automatically control the scale of c_1, c_2 and B.

10.5 Modified Models

Although the VMS model would not give values for B, c_1 and c_2 that are consistent with the image, the function u_{MS} (10.6) is usually a sensible piecewise-smooth approximation of the image, since the products Bc_1 and Bc_2 are finite. Once u_{MS} has converged, it is possible to manually rescale B, c_1, c_2 to finite values, but these are unknown quantities. The immediate question is: is it possible to incorporate constraints into the formulation in a reliable way, i.e. can we use information in the image to automatically restrict the scale of B, c_1 and c_2? Based on the observations of the previous section, it is natural to approach this problem in two ways. First, by explicitly restricting the size of the intensity constants, c_1 and c_2, and second, by controlling the scale of B.

10.5.1 Method 1

If $\tilde{c}_1$ and $\tilde{c}_2$ are good approximations of the 'correct' values of c_1 and c_2, then we can incorporate them into the formulation. It is unnecessary to constrain both intensity constants, and so we proceed by restricting c_1 by adjusting the formulation as follows, by including $\tilde{c}_1$:

$$E_1^{vms}(c_1, c_2, B) = \lambda \int_\Omega \left((z - Bc_1)^2 \chi_1 + (z - Bc_2)^2 \chi_2 \right) \, dx + v|\Gamma|$$

$$+ \mu \int_\Omega |\nabla B|^2 \, dx + \gamma_1 ||c_1 - \tilde{c}_1||^2, \tag{10.8}$$

where γ_1 is a positive parameter. The introduction of this constraint to VMS alters the minimisation of the intensity constants discussed in Sect. 10.4. Minimising (10.8) with respect to c_1 gives:

$$c_1 = \frac{\gamma_1 \tilde{c}_1 + \lambda \int_\Omega z(x) B(x) \chi_1 \, dx}{\gamma_1 + \lambda \int_\Omega B^2(x) \chi_1 \, dx}. \tag{10.9}$$

10.5.2 Method 2

Secondly, we tackle the problem of the scale of the bias field, B. With VMS, B is encouraged to be close to 0, which leads to the lack of convergence for c_1 and c_2. To prevent this we introduce a new term that encourages B to be close to a positive constant. The new formulation is given as follows:

$$E_2^{vms}(c_1, c_2, B) = \lambda \int_\Omega \left((z - Bc_1)^2 \chi_1 + (z - Bc_2)^2 \chi_2 \right) \, dx + v|\Gamma|$$

$$+ \mu \int_\Omega |\nabla B|^2 \, dx + \gamma_2 \int_\Omega (B - s)^2 \, dx, \tag{10.10}$$

where s, γ_2 are positive parameters. Here, with $\gamma_2 = 0$, clearly we just have VMS. If $s = 1$, and $\gamma_2 \to \infty$ we have the CV formulation (10.2). We intend to use the framework of VMS to approximate the exact minimiser of (10.10) for B. With this in mind, the previous formulation is equivalent to

$$E_2^{vms}(c_1, c_2, B) = \lambda \int_\Omega \left(\left[(z - Bc_1)^2 + \tfrac{\gamma_2}{\lambda}(B - s)^2 \right] \chi_1 \right.$$

$$+ \left[(z - Bc_2)^2 + \tfrac{\gamma_2}{\lambda}(B - s)^2 \right] \chi_2 \Big) \, dx$$

$$+ \mu \int_\Omega |\nabla B|^2 \, dx + v|\Gamma|,$$

where the new constraint has been incorporated into the fitting term. We can reformulate this as follows, first looking at the χ_1 term:

$$
\left[(z - Bc_1)^2 + \tfrac{\gamma_2}{\lambda}(B - s)^2\right] = B^2\left(c_1^2 + \tilde{\gamma}\right) - 2B\left(c_1 z + \tilde{\gamma} s\right) + \left(z^2 + \tilde{\gamma} s^2\right)
$$

$$
= \left(c_1^2 + \tilde{\gamma}\right)\left[B - \frac{c_1 z + \tilde{\gamma} s}{c_1^2 + \tilde{\gamma}}\right]^2 + f_1(z, c_1, s, \tilde{\gamma})
$$

$$
= \left[\frac{c_1 z + \tilde{\gamma} s}{\sqrt{c_1^2 + \tilde{\gamma}}} - B\sqrt{c_1^2 + \tilde{\gamma}}\right]^2 + f_1(z, c_1, s, \tilde{\gamma}),
$$

where $\tilde{\gamma} = \frac{\gamma_2}{\lambda}$ and $f_1(z, c_1, s, \tilde{\gamma}) = \frac{z^2 + \tilde{\gamma} s^2}{c_1^2 + \tilde{\gamma}}$. In a similar way, for the χ_2 term:

$$
\left[(z - Bc_2)^2 + \tfrac{\gamma_2}{\lambda}(B - s)^2\right] = \left[\frac{c_2 z + \tilde{\gamma} s}{\sqrt{c_2^2 + \tilde{\gamma}}} - B\sqrt{c_2^2 + \tilde{\gamma}}\right]^2 + f_2(z, c_2, s, \tilde{\gamma}),
$$

where $f_2(z, c_2, s, \tilde{\gamma}) = \frac{z^2 + \tilde{\gamma} s^2}{c_2^2 + \tilde{\gamma}}$. Therefore (10.10) is equivalent to

$$
E_2^{vms}(c_1, c_2, B) = \lambda \int_{\Omega}\left(\left[\frac{c_1 z + \tilde{\gamma} s}{\sqrt{c_1^2 + \tilde{\gamma}}} - B\sqrt{c_1^2 + \tilde{\gamma}}\right]^2 \chi_1\right.
$$

$$
+ \left.\left[\frac{c_2 z + \tilde{\gamma} s}{\sqrt{c_2^2 + \tilde{\gamma}}} - B\sqrt{c_2^2 + \tilde{\gamma}}\right]^2 \chi_2\right) dx
$$

$$
+ \int_{\Omega} f_1(z, c_1, s, \tilde{\gamma})\chi_1\, dx + \int_{\Omega} f_2(z, c_2, s, \tilde{\gamma})\chi_2\, dx
$$

$$
+ \mu \int_{\Omega} |\nabla B|^2\, dx + v|\Gamma|.
$$

Minimising $E_2^{vms}(c_1, c_2, B)$ with respect to B is given by

$$
\min_{B}\left\{\lambda \int_{\Omega}\left([\bar{z}_1 - B\bar{c}_1]^2\chi_1 + [\bar{z}_2 - B\bar{c}_2]^2\chi_2\right)dx + \mu \int_{\Omega}|\nabla B|^2\, dx\right\}, \qquad (10.11)
$$

since $f_1(z, c_1, \mathbf{s}, \tilde{\gamma}), f_2(z, c_2, \mathbf{s}, \tilde{\gamma})$, and $|\Gamma|$ are not dependent on B. Here

$$\bar{z}_1 = \frac{c_1 z + \tilde{\gamma}\mathbf{s}}{\sqrt{c_1^2 + \tilde{\gamma}}}, \quad \bar{c}_1 = \sqrt{c_1^2 + \tilde{\gamma}}, \quad \bar{z}_2 = \frac{c_2 z + \tilde{\gamma}\mathbf{s}}{\sqrt{c_2^2 + \tilde{\gamma}}}, \text{ and } \bar{c}_2 = \sqrt{c_2^2 + \tilde{\gamma}}.$$

In the same way as VMS, we can approximate the exact minimiser of (10.11):

$$B = \frac{\bar{c}_1 \bar{z}_1 \chi_1 + \bar{c}_2 \bar{z}_2 \chi_2}{\bar{c}_1^2 \chi_1 + \bar{c}_2^2 \chi_2} * G. \tag{10.12}$$

10.6 Results

An important aspect of the success of each method is how much the constraints affect the final segmentation. With this in mind, we can treat the result computed by VMS as the 'ground truth' and quantifiably measure the solution of each method against this using the Tannimoto Coefficient [9]:

$$TC = \frac{N(GT \cap \Omega_1^*)}{N(GT \cup \Omega_1^*)}, \tag{10.13}$$

where $N(\cdot)$ is the number of pixels in the enclosed region, GT is the result computed by VMS, and Ω_1^* is the result computed with additional constraints. The final intensity constants computed by each method are c_1^* and c_2^*. Another aspect of the success of each method is what these values are; we can check whether the computed values are feasible, i.e. $c_1, c_2 \in [0, 1]$, whilst maintaining the quality of the segmentation.

For VMS Method 1 (M1), we test two examples (Images 1 and 2, see Fig. 10.4), both used in [8]. In Tables 10.1 and 10.2, we present results for variations of the initial intensity approximation, $\tilde{c}_1$, and the constraint parameter, γ_1, in terms of how accurately these values can reproduce the results of VMS. In addition, Figs. 10.5 and 10.6 contain example results for M1, demonstrating the convergence of c_1 and c_2. For VMS Method 2 (M2), we test four examples (Images 1–4, see

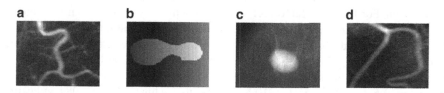

Fig. 10.4 Images tested with M1 and M2. (**a**) Image 1. (**b**) Image 2. (**c**) Image 3. (**d**) Image 4

Table 10.1 Result Sets for Method 1, Image 1 (bold results shown in Fig. 10.5)

$\tilde{c}_1$		Constraint parameter, γ_1							
		1	10	10^2	10^3	10^4	10^5	10^6	10^7
0.2	c_1^*	3.773	4.928	1.625	0.591	0.282	0.212	0.201	0.200
	c_2^*	2.729	8.017	2.628	0.958	0.450	0.356	0.338	0.337
	TC	0.999	0.001	0.001	0.001	0.001	0.001	0.001	0.001
0.3	c_1^*	3.763	2.707	1.221	0.526	0.336	0.304	0.300	0.300
	c_2^*	2.721	1.994	0.915	0.397	0.253	0.222	0.226	0.226
	TC	0.999	0.996	0.991	0.964	0.962	0.962	0.962	0.962
0.4	c_1^*	3.750	2.723	1.277	0.599	0.428	0.403	0.400	0.400
	c_2^*	2.711	2.004	0.957	0.452	0.323	0.304	0.302	0.302
	TC	0.999	0.996	0.991	0.965	0.962	0.962	0.962	0.962
0.5	c_1^*	3.679	2.753	1.339	0.678	**0.528**	0.502	0.500	0.500
	c_2^*	2.659	2.025	1.004	0.510	**0.393**	0.379	0.377	0.377
	TC	0.999	1.000	0.991	0.986	**0.984**	0.962	0.962	0.962
0.6	c_1^*	3.686	2.780	1.403	0.759	0.620	0.602	0.600	0.600
	c_2^*	2.664	2.042	1.052	0.571	0.466	0.454	0.452	0.452
	TC	0.999	1.000	0.991	0.986	0.984	0.962	0.962	0.962
0.7	c_1^*	3.691	2.812	1.470	0.843	0.717	0.702	0.700	0.700
	c_2^*	2.667	2.067	1.103	0.634	0.539	0.529	0.528	0.528
	TC	1.000	1.000	0.992	0.986	0.984	0.962	0.962	0.962
0.8	c_1^*	3.693	2.847	1.540	0.931	0.814	0.802	0.800	0.800
	c_2^*	2.668	2.092	1.155	0.699	0.614	0.604	0.603	0.603
	TC	1.000	1.000	0.992	0.975	0.962	0.962	0.969	0.969

The TC row demonstrates how this method compares to the VMS results, while rows for c_1, c_2 suggest reliance on $\tilde{c}_1$

Fig. 10.4), all used in [8]. In Fig. 10.7 we present examples of M2 for Images 1–4. The constraint parameter is fixed at $s = 1$, with γ_2 depending on each case. Below we discuss the results for each method in detail.

For M1 we test two challenging examples. Image 1 is an example of vessel segmentation, from Fig. 6 of [8]. It contains intensity variation in the foreground, and background artefacts of a similar intensity to the object, which makes it a difficult example. The VMS result for this case shows a minor inaccuracy in the sense that one artefact is included in the segmented region, which we hope to improve on. For Image 1, Table 10.1 shows that a high value of TC can be achieved except for very small values of the initial intensity approximation, i.e. $\tilde{c}_1 = 0.2$. As γ_1 increases the additional constraint of M1 (10.8) is enforced exactly, as expected. With $\gamma > 10^3$, $c_1^*, c_2^* \in [0, 1]$, suggesting that γ_1 should be higher than this to be consistent with the image. For Image 1, we have shown an example of a result for $\tilde{c}_1 = 0.5, \gamma_1 = 10^5$ in Fig. 10.5. We see that whilst $TC = 0.984$ indicating a slight variation from the result computed from VMS, we still get a good result. However, as seen in Fig. 10.3 and Fig. 6 of [8] the result we took as a 'ground truth', as mentioned before, does contain an inaccuracy as it includes a slight artefact in

Table 10.2 Result Sets for Method 1, Image 2 (bold results shown in Fig. 10.6)

$\tilde{c}_1$		Constraint parameter, γ_1							
		1	10	10^2	10^3	10^4	10^5	10^6	10^7
0.2	c_1^*	1.576	1.547	1.331	0.730	0.333	0.250	0.205	0.200
	c_2^*	1.406	1.383	1.203	0.674	0.309	0.318	0.246	0.235
	TC	1.000	1.000	1.000	0.999	0.998	0.998	0.059	0.141
0.3	c_1^*	1.576	1.552	1.358	0.790	0.409	0.314	0.301	0.300
	c_2^*	1.407	1.387	1.226	0.729	0.379	0.292	0.279	0.278
	TC	1.000	1.000	1.000	0.999	0.998	0.998	0.673	0.659
0.4	c_1^*	1.577	1.556	1.386	0.855	0.491	0.411	0.401	0.400
	c_2^*	1.407	1.390	1.251	0.789	0.455	0.381	0.372	0.371
	TC	1.000	1.000	1.000	1.000	0.998	0.998	0.997	0.997
0.5	c_1^*	1.577	1.560	1.416	0.924	**0.577**	0.509	0.501	0.500
	c_2^*	1.407	1.394	1.277	0.853	**0.535**	0.472	0.464	0.464
	TC	1.000	1.000	1.000	0.999	**0.998**	0.997	0.890	0.896
0.6	c_1^*	1.578	1.564	1.447	0.996	0.667	0.605	0.601	0.600
	c_2^*	1.408	1.397	1.304	0.920	0.619	0.561	0.557	0.557
	TC	1.000	1.000	1.000	0.999	0.998	0.896	0.885	0.885
0.7	c_1^*	1.578	1.567	1.478	1.071	0.759	0.704	0.700	0.700
	c_2^*	1.408	1.400	1.332	0.990	0.704	0.653	0.649	0.648
	TC	1.000	1.000	1.000	0.999	0.998	0.885	0.777	0.695
0.8	c_1^*	1.578	1.571	1.510	1.149	0.852	0.803	0.800	0.800
	c_2^*	1.409	1.403	1.359	1.062	0.791	0.754	0.739	0.739
	TC	1.000	1.000	1.000	0.999	0.991	0.886	0.557	0.557

The TC row demonstrates how this method compares to the VMS results, while rows for c_1, c_2 suggest reliance on $\tilde{c}_1$

Fig. 10.5 Results for Image 1 (with $\gamma_1 = 10^5$ and $\tilde{c}_1 = 0.5$). *Top left* shows progression of c_1 values (vertical axis) against iterations (horizontal axis)—c_2 values demonstrate similar behaviour. *Top right* is the bias field B. *Bottom left* is the result for VMS Method 1 ($TC = 0.962$), and *bottom right* is original VMS, demonstrating our method can improve quality. (**a**) c_1 Progression. (**b**) Bias field. (**c**) New VMS, Method 1 Contour. (**d**) Old VMS Contour

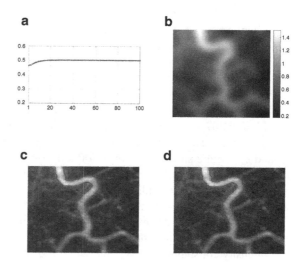

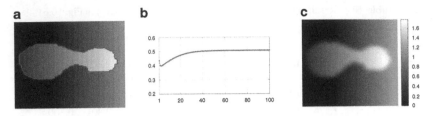

Fig. 10.6 Results for Image 2 (with $\gamma = 10^5$ and $\tilde{c}_1 = 0.5$). From *left to right*: result for VMS Method 1 ($TC = 0.997$), progression of c_1 values (vertical axis) against iterations (horizontal axis)—c_2 values demonstrate similar behaviour, and bias field B. (**a**) VMS, Method 1 Contour. (**b**) c_1 Progression. (**c**) Bias field, B

the background. This is the best result computed for VMS, with many variations of λ and σ, and agrees with the results of D. Chen et al. [8]. This demonstrates that M1 can actually improve the results for difficult examples, such as Image 1. Image 2, from Fig. 1 of [8], contains very large intensity variation in both the foreground and background. The VMS result for this case is good, but the challenge is how robust Method 1 and 2 are to the additional parameters. For Image 2, we see that for high values of γ_1 the final result, Ω_1^*, can deviate from the result computed by VMS, except for good selections of $\tilde{c}_1$, i.e. $\tilde{c}_1 = 0.4, 0.5$. With $\gamma_1 > 10^4$, $c_1^*, c_2^* \in [0, 1]$. This demonstrates that for this particularly challenging case of intensity inhomogeneity, which has high intensity variation in the foreground and background the success of M1 can be more sensitive to the additional constraint than in Image 1. However, Fig. 10.6 shows the desired convergence of c_1 and c_2, allowing us to automatically recover a feasible 'true' image for appropriate parameters.

For M2 we test four examples (Images 1–4), all used in D. Chen et al. [8]. In Fig. 10.7, we present the results for each case. We set $\mathbf{s} = 1$, and vary the constraint parameter for each case. Given this value, for an $n \times n$ image, we expect $||B||_F \approx 128$ with the addition of the M2 constraint. For Image 1 $\gamma_2 = 0.1$, and the intensity constants converge to $c_1^* = 0.352, c_2^* = 0.240$. The bias field, B, also converges and we compute $||B||_F = 129.8$. This result is of a similar quality to VMS, shown in Fig. 10.3, but of slightly worse quality to the result for M1, shown in Fig. 10.5. For Image 2 $\gamma_2 = 0.2$, and the intensity constants converge to $c_1^* = 0.807, c_2^* = 0.660$. The bias field, B, also converges and we compute $||B||_F = 67.0$. However, row 2 of Fig. 10.7 demonstrates that the convergence of c_1 and c_2 is quite slow, taking over 500 iterations which is much more than for the convergence of u_{MS} in VMS. Image 3 is an ultrasound image, containing intensity variation in the background. For this example, $\gamma_2 = 0.1$, and the intensity constants converge to $c_1^* = 0.446, c_2^* = 0.383$. The bias field, B, also converges and we compute $||B||_F = 120.9$. Image 4 is another example of vessel segmentation, where intensity varies smoothly throughout the vessel. For this example, $\gamma_2 = 0.1$, and the intensity constants converge to $c_1^* = 0.352, c_2^* = 0.239$. The bias field, B, also converges and we compute $||B||_F = 122.2$. We see results of comparable quality to rows 3 and 4 in Fig. 10.7, in Figs. 9 and 4 of [8], respectively, except that there is no convergence for c_1, c_2 and B. This demonstrates a clear improvement for these

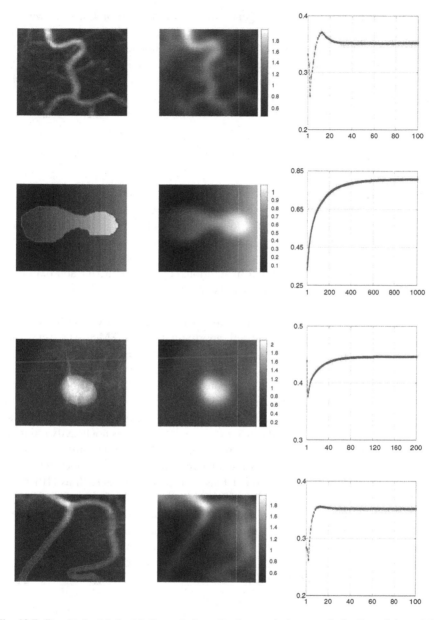

Fig. 10.7 Results for Method 2. Rows 1–4 are for Images 1–4, respectively. From *left to right*: result for VMS Method 2, bias field B, and the progression of c_1 values (vertical axis) against iterations (horizontal axis)—c_2 values demonstrate similar behaviour

examples. Despite the slow convergence of M2 in the case of Image 2, we have fast convergence in the other examples, meaning the additional constraint of M2 generally doesn't slow down the computation of a solution. Also, the results are not sensitive to the constraint parameter, γ_2. For Image 2, it was adjusted to 0.2, but for all other cases it was set at 0.1, and for all examples, $c_1^*, c_2^* \in [0, 1]$, showing that M2 produces results consistent with the image.

10.7 Conclusions

In this paper, we have introduced two modified VMS models that ensure the separate convergence of all variables minimised. VMS Method 1, restricting the growth of one intensity constant, allows for an intuitive control on the intensity of the desired object, whilst ensuring the convergence of each variable. VMS Method 2, restricting the scale of the bias field, B, is a simple adjustment of the original formulation of D. Chen et al. [8]. We have incorporated an additional constraint into the approximation of B. We see a good standard of results in four examples, where all variables converge separately.

We suggest that both modified models have advantages over VMS. In cases where the intensity constants can be well approximated, VMS Method 1 is a natural choice. It has demonstrated that it can achieve very good results, even improving on the results of D. Chen et al. [8] in a difficult example. However, some sensitivity to the additional parameter, γ_1 can limit its potential efficacy. The additional constraint of VMS Method 2 can reliably control the scale of the bias field, B, thus ensuring the convergence of all variables, and is therefore an improvement on VMS. It is a simple and effective addition to the original formulation, and is not heavily reliant on the additional parameters, and so does not significantly affect usability. In general, VMS Method 2 is the superior modification due to its flexibility; it does not require any prior knowledge. There exist other bias correction models such as [10] which could be studied and compared to VMS in future works.

Finally it should be remarked that while variational models are generally powerful, there are issues to be overcome: (a) Non-complexity—it may take a long time to completely resolve this problem; (b) Slow convergence—this is a long standing issue with nonlinear optimisation or PDEs; in some cases, developing a fast multigrid method is possible [2].

References

1. M. Ahmed, S. Yamany, N. Mohamed, A. Farag, T. Moriarty, A modified fuzzy c-means algorithm for bias field estimation and segmentation of mri data. IEEE Trans. Med. Imaging **21**(3), 193–199 (2002)
2. N. Badshah, K. Chen, On two multigrid algorithms for modelling variational multiphase image segmentation. IEEE Trans. Image Process. **18**, 1097–1106 (2009)

3. T. Brox, D. Cremers, On local region models and a statistical interpretation of the piecewise smooth MS functional. Int. J. Comput. Vis. **84**(2), 184–193 (2009)
4. V. Caselles, R. Kimmel, G. Sapiro, Geodesic active contours. Int. J. Comput. Vis. **22**(1), 61–79 (1997)
5. A. Chambolle, An algorithm for total variation minimization and applications. J. Math. Imaging Vis. **20**, 89–97 (2004)
6. T.F. Chan, L. Vese, Active contours without edges. IEEE Trans. Image Process. **10**(2), 266–277 (2001)
7. T.F. Chan, S. Esedoglu, M. Nikilova, Algorithms for finding global minimizers of image segmentation and denoising models. SIAM J. Appl. Math. **66**, 1932–1648 (2006)
8. D. Chen, M. Yang, L. Cohen, Global minimum for a variant Mumford-Shah model with application to medical image segmentation. Comput. Methods Biomech. Biomed. Eng. Imaging Vis. **1**(1), 48–60 (2013)
9. W. Crum, O. Camara, D. Hill, Generalized overlap measures for evaluation and validation in medical image analysis. IEEE Trans. Med. Imaging **25**(11), 1451–1461 (2006)
10. Y. Duan, H. Chang, W. Huang, J. Zhou, Simultaneous bias correction and image segmentation via L0 regularized Mumford-Shah model, in *ICIP 2014*
11. M. Jung, G. Peyre, L. Cohen, Nonlocal active contours. SIAM J. Imag. Sci. **5**(3), 1022–1054 (2012)
12. M. Kass, A. Witkin, D. Terzopoulos, Snakes: active contour models. Int. J. Comput. Vis. **1**(4), 321–331 (1988)
13. S. Lanktona, A. Tannenbaum, Localizing region-based active contours. IEEE Trans. Image Process. **17**(11), 2029–2039 (2008)
14. C. Li, C. Kao, J. Gore, Z. Ding, Minimization of region-scalable fitting energy for image segmentation. IEEE Trans. Image Process. **17**(10), 1940–1949 (2008)
15. C. Li, Z. Rui Huang, J. Gatenby, D. Metaxas, J. Gore, Global minimum for a variant Mumford-Shah model with application to medical image segmentation. Comput. Methods Biomech. Biomed. Eng. **20**(7), 2007–2016 (2011)
16. F. Li, M. Ng, C. Li, Variational fuzzy Mumford-Shah model for image segmentation. SIAM J. Appl. Math. **70**(7), 2750–2770 (2010)
17. D. Mumford, J. Shah, Optimal approximation by piecewise smooth functions and associated variational problems. Commun. Pure Appl. Math. **42**, 577–685 (1989)
18. M. Nielsen, L. Florack, R. Deriche, Regularization and scale space. Technical Report, INRIA, 1994
19. S. Osher, J. Sethian, Fronts propagating with curvature-dependent speed: algorithms based on Hamilton-Jacobi formulations. J. Comput. Phys. **79**(1), 12–49 (1988)
20. K. Sun, Z. Chen, S. Jiang, Local morphology fitting active contour for automatic vascular segmentation. IEEE Trans. Biomed. Eng. **59**(2), 464–473 (2012)
21. A. Tsai, A. Yezzi, A. Willsky, Curve evolution implementation of the Mumford-Shah functional for image segmentation, denoising, interpolation, and magnification. IEEE Trans. Image Process. **10**(8), 1169–1186 (2001)
22. H. Zhao, T.F. Chan, B. Merriman, S. Osher, A variational level set approach to multiphase motion. J. Comput. Phys. **127**(1), 179–195 (1996)

Chapter 11
An On-Line Strategy of Groups Evacuation from a Convex Region in the Plane

Bo Jiang, Yuan Liu, Hao Zhang, and Xuehou Tan

Abstract This paper considers the problem faced by people who must evacuate an area as quickly as possible. The evacuees are divided into multiple groups share information completely. We seek strategies that achieve a bounded ratio of evacuation path length, without any knowledge of boundary information. We restrict the affected area to a convex region in the plane, and present a $4\sqrt{2}$-competitive strategy for n groups of evacuees ($n \geq 3$). The performance of the strategy is evaluated with case studies.

11.1 Introduction

Motivated by its relations to the well known searching and evacuation problem, recently much attention has been devoted to the problem of how to efficiently evacuate an affected area in an emergency. In this paper, suppose that the evacuees are divided into multiple groups, at different locations in the affected area. They don't know any boundary information of the affected area, but can share information with each other during the evacuation. We seek strategies to evacuate the affected people from the dangerous region as quickly as possible. We call the problem discussed in this paper the multi-source points evacuation problem, since the evacuees start to evacuate from several different initial positions in the affected area.

An on-line strategy is usually evaluated by a competitive ratio. Let P denote an affected area, restricted to a convex region in the plane. The evacuees are modelled as the points inside P, whose initial positions are at different locations. When our strategy is used to evacuate P, without any boundary information we let the evacuation time t_{tactic} denote the cost of the tactic. The evacuation time t_{tactic} is the

B. Jiang (✉)
Dalian Maritime University, Dalian, China
e-mail: bojiang@dlmu.edu.cn

Y. Liu • H. Zhang
School of Information Science and Technology, Dalian Maritime University, Dalian, China

X. Tan
School of Information Science and Technology, Tokai University, Tokyo, Japan

© Springer International Publishing Switzerland 2015
L.M. Chen et al., *Mathematical Problems in Data Science*,
DOI 10.1007/978-3-319-25127-1_11

period from the beginning of the evacuation to the moment that all people evacuees successfully evacuate the affected area. Let t_{opt} denote the shortest evacuation time required to evacuate from P in the case that the evacuees know the boundary information of P. If the boundary information of the affected area is known, they can reach the boundary in the shortest time or distance. The competitive ratio of an on-line strategy is defined as the upper bound of t_{tactic} / t_{opt}.

Previous Work The evacuation problem has been extensively studied. Chen et al. [1] used the methods of system simulation to compare the evacuation efficiency of three different road networks. Lu et al. [2] and Shekhar et al. [3] studied the shortest path algorithm of evacuation with the consideration of capacity constraints and increasing the number of people in time and space. Berman [4] surveyed the problems of online searching and navigation. Burgard [5] considered the evacuate ratio with the cost of strategy as the time spent, instead of path lengths.

To analyze the evacuation strategy under complete information on the boundary, previous studies mainly focus on details such as flow and other constraints. In a recent work [6], Xu et al. considered a new situation in which the evacuees don't know the boundary information of the affected area. This actually occurs in an emergency, as the affected region is not known to the evacuees in some cases. For the convex region in the plane, Xu et al. [6] gave an evacuation strategy for $n(n \geq 3)$ groups of the evacuees with a competitive ratio of $3/cos(\pi/n)$, provided that the evacuees can communicate with each other during the evacuation. For $n = 1$, Wei et al. [8] gave an evacuation strategy with a competitive ratio of 19.64. But, for $n = 2$, the problem is unsolved. For the case $n = 3$, Wei et al. [7] improved the competitive ratio to $2 + 2\sqrt{3}$. The evacuation strategies for grid networks are also studied in [6].

Our Work In the strategy of Xu et al. [6], the evacuees are modelled as the points inside P, whose initial positions are the same. In this paper, we further consider the case where the initial positions are different, meaning our model is more practical. Also, we assume that the evacuees can communicate with each other and move at the unit speed during the evacuation. Our main result is a $4\sqrt{2}$-*competitive* strategy for the evacuation of several groups, and thus solves the multi-source points evacuation problem.

The rest of this paper is organized as follows. In Sect. 11.2, we give some basic definitions relevant to this paper. In Sect. 11.3, we present an on-line strategy to solve the multi-source points evacuation problem and give its competitive ratio. In Sect. 11.4, we conclude the paper with a discussion of further research and open questions.

11.2 Preliminaries

In this paper, we define a convex polygon as a closed polygonal chain with all interior angles are less than π. Let P be a convex polygon. Let $O = \{O_1, O_2, \ldots, O_n\}$

denote a sequence of initial locations. The evacuees starting at $O_i, 1 \leq i \leq n$, don't know the boundary information. The evacuees at an initial location converge to a group, and their goal is to leave from P as soon as possible. Let $G = \{G_1, G_2, \ldots, G_n\}$ denote the divided groups of evacuees.

Successful evacuation requires that all evacuees reach the boundary of the affected area. We suppose that evacuees move at the unit speed during the evacuation. So, the cost of our strategy is the path length of the last group to evacuate. Let d_{tactic} denote the distance this group walks using our strategy. Let d_{opt} denote the largest distance among all groups to the boundary of P, provided that the boundary of P is known in advance. The performance of a strategy is evaluated by a competitive ratio k, is defined above. Then, we get that $k = t_{tactic}/t_{opt} = d_{tactic}/d_{opt}$. Our objective is to minimize the competitive ratio k.

We make some assumptions about the evacuation in this paper. First, the evacuees, at the initial locations $O_1, O_2, \ldots, O_n$ in P, don't know the boundary information of P, but they can share information all the time. Second, the evacuees move at the unit speed during the evacuation. Third, the longest distance among the original locations is much less than d_{opt} (i.e., $<< d_{opt}$).

11.3   The Multi-Source Points Evacuation Problem

In this section, we study tactics in evacuation problem. In reality, the evacuees have several grouping options. One is that all the evacuees converge to a group. Another is that the evacuees are divided into several groups, and choose different paths to escape the affected area with information sharing between them. The evacuation strategy of n groups ($n \geq 3$) had been studied in [6], provided that the communication between different groups is always possible. Our strategy also allows evacuees to divide into several groups to escape the affected area with information sharing between them. Moreover, groups of evacuees can start at different locations.

Our Tactic Is Described as Follows

Step 1:   The evacuees at an original location O_i are considered as a group, and we denote by G_i the evacuees group at O_i.

Step 2:   Find the convex hull of the original locations $O_1, O_2 \ldots O_n$. Move the evacuees from the locations, which are not on the vertex of the hull, to the closest vertex on the hull. In Fig. 11.1, G_2 is moved from O_2 to O_4, and G_3 is moved from O_3 to O_6. Then renumber the original locations on the hull, see Fig. 11.2. From the third assumption above, the distances of moving the groups to the convex hull can be ignored.

Step 3:   Design evacuation paths for all groups at the renumbered original locations on the hull. Let R_i denote the path starting from the location O_i. The paths are radials and designed as follows. First, the path for the evacuees at the first location O_1 is the bisector of the interior angle $\angle O_1$. The direction of the path is outward. Second, the path for the evacuees at the other location O_i, beside O_1, is

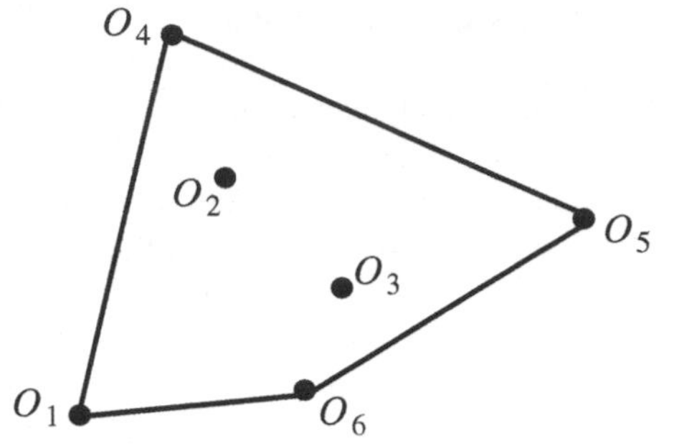

Fig. 11.1 The convex hull of
all initial locations

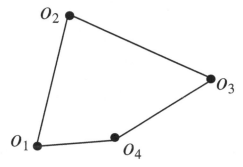

Fig. 11.2 Only the locations
on the convex hull are
renumbered

Fig. 11.3 Illustrating the
evacuation paths

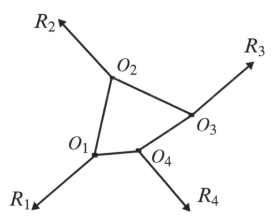

the path from O_{i-1} rotated $2\pi/n$ (n is the number of evacuees groups) clockwise,
$i = 2, \ldots, n$. See Fig. 11.3, where $n = 4$. The path starting from O_1 is a radial,
which is on the bisector of interior angle $\angle O_1$. R_2 is R_1 rotated $2\pi/4$ clockwise.
Also, R_3 is R_2 rotated $2\pi/4$ clockwise, and $R4$ is $R3$ rotated $2\pi/4$ clockwise.

Step 4:   When one group G_i, reaches the boundary of P first, other groups which
           have not arrived at the boundary of P change their directions and move towards
           the boundary point that had been reached by G_i. Let L_j denote the location of G_j
           when the first group reaches the boundary of P, at L_i, see Fig. 11.4. The group

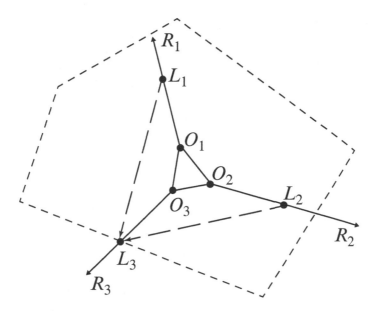

Fig. 11.4 An instance of three groups evacuation problem

G_3 starting from O_3 reaches the boundary of P first. When G_3 arrives at L_3 on the boundary of P, the group G_1, which arrives at L_1, and the group G_2, which arrives at L_2, change their directions towards L_3 and continue to move. Finally, all the evacuees successfully evacuate the affected area P through the point L_3.

Theorem 11.1. *The evacuate ratio of our tactic with* n(n $\geq$ 3) *groups is no more than* $4\sqrt{2}$.

Proof. Let O_t, $1 \leq t \leq n$, denote the initial location of the group G_t in P and L_t denote the current point of G_t, when one of $G_1, \ldots, G_n$ reaches the boundary of P. More specifically, suppose that L_i is on the boundary of P and L_j is any other point, G_j changes its direction and moves from L_j to L_i ($j \neq i$). Let d denote the distance which the group first walks to find the boundary, then the total distance moved by G_j is $d + |L_iL_j|$. The original locations O_i and O_j may be on the same side of segment L_iL_j, or on the different side of segment L_iL_j, see Fig. 11.5. Whether they are on the same or the different side, we can get $|L_iL_j| \leq |O_iL_i| + |O_iO_j| + |O_jL_j|$. In other words, $|L_iL_j| \leq 3d$. So, we have that $d_{tactic} = \max\{|L_iL_j|\} + d$, then $d_{tactic} \leq 4d$.

Now, we compute d_{opt} as follows. Let P' denote the convex polygon of the vertices $L_1, L_2, \ldots, L_n$. Clearly, P' is completely contained in P, see Fig. 11.6, where $n = 3$.

Suppose the boundary of the affected area is the polygon P', then d_{opt} is the largest distance from the original locations to the boundary of P' among all groups. Therefore, d_{opt} can be obtained by computing the shortest distance from the original locations to the boundary of P'. There are four situations as follows.

Fig. 11.5 Different situations
for points O_i, O_j and segment
L_iL_j. (**a**) O_i and O_j are on the
same side of segment L_iL_j.
(**b**) O_i and O_j are on the
different side of segment L_iL_j

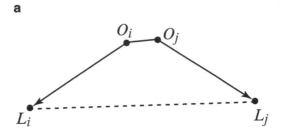

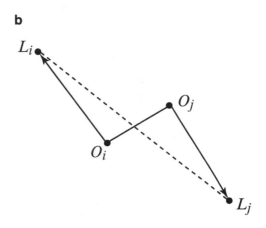

Fig. 11.6 Illustrating the
convex polygon $P'(n = 3)$

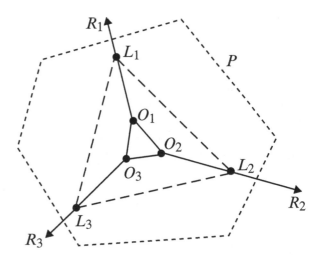

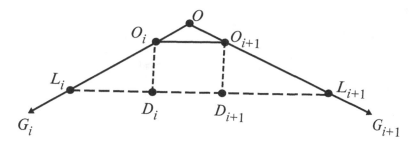

Fig. 11.7 The distance from source points to the boundary of P'(Case 1)

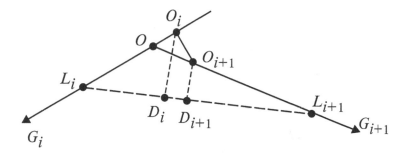

Fig. 11.8 The distance from source points to the boundary of P'(Case 2)

Case 1:   The intersection of line O_iL_i and $O_{i+1}L_{i+1}$ is not on the segment O_iL_i and $O_{i+1}L_{i+1}$ (Fig. 11.7). The foot of a perpendicular from O_i to line L_iL_{i+1} is D_i, which is on the segment L_iL_{i+1}. Also, the foot of a perpendicular from O_{i+1} to line L_iL_{i+1} is D_{i+1}, which is on the segment L_iL_{i+1}.

We know that $\angle L_iOL_{i+1} = 2\pi/n$ and $O_iL_i = O_{i+1}L_{i+1} = d$. If O_iO_{i+1} is parallel to L_iL_{i+1}, we get $O_iD_i = O_{i+1}D_{i+1} = d\cos(\pi/n)$. If O_iO_{i+1} is not parallel to L_iL_{i+1}, we get that $\min\{O_iD_i, O_{i+1}D_{i+1}\} < d\cos(\pi/n) < \max\{O_iD_i, O_{i+1}D_{i+1}\}$.

Case 2:   The intersection of line O_iL_i and $O_{i+1}L_{i+1}$ is on the segment O_iL_i or $O_{i+1}L_{i+1}$ (Fig. 11.8). The foot of a perpendicular from O_i to line L_iL_{i+1} is D_i, which is on the segment L_iL_{i+1}. Also, the foot of a perpendicular from O_{i+1} to line L_iL_{i+1} is D_{i+1}, which is on the segment L_iL_{i+1}.

We know that $\angle L_iOL_{i+1} = 2\pi/n$ and $O_iL_i = O_{i+1}L_{i+1} = d$. Then, $OL_i < OL_{i+1}$, $\angle OL_iL_{i+1} > \pi/2 - \pi/n > \angle OL_{i+1}L_i$. Because $O_iD_i = O_iL_i\sin\angle O_iL_iL_{i+1}$, we get that $O_iD_i = d\sin\angle O_iL_iL_{i+1} > d\sin(\pi/2 - \pi/n) = d\cos(\pi/n)$, that is, $O_iD_i > d\cos(\pi/n)$.

Case 3:   The intersection of line O_iL_i and $O_{i+1}L_{i+1}$ is not on the segment O_iL_i and $O_{i+1}L_{i+1}$ (Fig. 11.9). The foot of a perpendicular from O_i to line L_iL_{i+1} is D_i, which is not on the segment L_iL_{i+1}. But the foot of a perpendicular from O_{i+1} to line L_iL_{i+1} is D_{i+1}, which is on the segment L_iL_{i+1}.

Fig. 11.9 The distance from
source points to the boundary
of P'(Case 3)

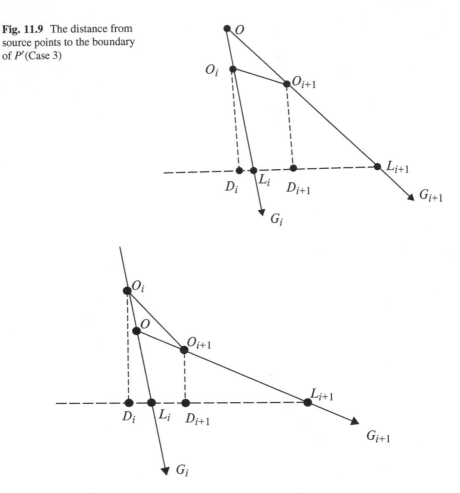

Fig. 11.10 The distance from source points to the boundary of P'(Case 4)

We know that $\angle L_i O L_{i+1} = 2\pi/n$ and $O_i L_i = O_{i+1} L_{i+1} = d$. See Fig. 11.9.
The shortest distance from O_i to the segment $L_i L_j$ is the segment $O_i L_i$. Then,
$O_i L_i = d > d \cos(\pi/n)$.

Case 4.   The intersection of line $O_i L_i$ and $O_{i+1} L_{i+1}$ is on the segment $O_i L_i$ or
$O_{i+1} L_{i+1}$ (Fig. 11.10). The foot of a perpendicular from O_i to line $L_i L_{i+1}$ is D_i,
which is not on the segment $L_i L_{i+1}$. But the foot of a perpendicular from O_{i+1} to
line $L_i L_{i+1}$ is D_{i+1}, which is on the segment $L_i L_{i+1}$.

See Fig. 11.10. The shortest distance from O_i to the segment $L_i L_j$ is the segment
$O_i L_i$. Then, $O_i L_i = d > d \cos(\pi/n)$.

As discussed above, Cases $1 \sim 4$ enumerate all possible situations, so we know
that at least one of the distances, which are from the original locations to the
boundary between two adjacent paths, is longer than $d \cos(\pi/n)$. According to the

definition of d_{opt}, $d_{opt} \geq d \cos(\pi/n)$.

$$k = t_{tactic}/t_{opt} = d_{tatic}/d_{opt} \leq \frac{4d}{d \cos \frac{\pi}{n}} = \frac{4}{\cos \frac{\pi}{n}} \quad (11.1)$$

Let $x = \cos \frac{\pi}{n}$, $n > 3$, $x \in [\frac{\sqrt{2}}{2}, 1)$. Then, $k \leq f(x) = \frac{4}{x}$. $f(x)$ is a decreasing function, so $k \leq f(\frac{\sqrt{2}}{2}) = 4\sqrt{2}$.

When $n = 3, k \leq 4\sqrt{2}$ also holds since in the case of evacuating from the same initial position the competitive ratio is $k \leq 2 + 2\sqrt{3}$ [7], which is at least the path length moved by evacuees in multi-source points evacuation problem.

The proof is thus complete.

11.4   Concluding Remarks

In this paper, we study the problem of finding efficient strategies for evacuating a convex region without any boundary information as soon as possible. We consider the problem for several groups evacuation, and present a $4\sqrt{2}-competitive$ strategy to solve multi-source points evacuation problem.

If we release the communication restriction, how can the evacuees escape from the affected area? In this situation, Wei et al. [8] have solved the problem of only one group of the evacuees to escape from the affected area and have presented a 19.64-competitive strategy. The main idea of their strategy is to iteratively follow a sequence of semicircles, whose centers are all the same, such that the radius of a semicircle is twice that of the previous one in the sequence. It can be considered as the planar generalization of the *doubling strategy* [9]. Let $|OPT(P)|$ denote the shortest distance between the evacuees and the boundary of P. The very first radius is assumed to be on the horizontal line L and of length one, which is assumed to be small, as compared to $|OPT(P)|$. Two consecutive semicircles are simply connected by a line segment. An instance of the strategy is shown in Fig. 11.11, where O denotes the starting position of the evacuees' group, and A_i, B_i the points of L intersecting the ith half-circle in the strategy. OA_1 is the first radius, $|OA_1| = 1$. OA_2 is the second radius, $|OA_2| = 2$. OA_3 is the third radius, $|OA_3| = 4$, and so on. $|OPT(P)|$ is $|OI|$. The strategy is called spiral evacuation strategy described as follows.

Step 1:   Make a directed line L, with arbitrary direction, which passes through O, as shown in Fig. 11.11. Let $Z = \{Z1, Z2\} = \{+\overline{L}, -\overline{L}\}$ denote the direction set defined by L, where $+\overline{L}$ is the direction of L, and $-\overline{L}$ is the direction opposite to $+\overline{L}$. Let j denote the sequence number of Z and k denote the sequence number of the set of half circles. At the start, let $j = 1, k = 1$ $(j, k \in N)$ and G, the group of the evacuees, begins at point O.

Fig. 11.11 The strategy to solve one group evacuation problem

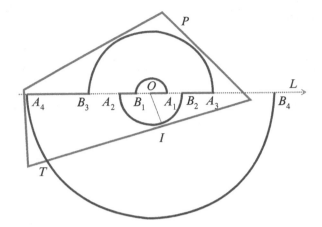

Step 2:   Make a segment $\overline{OA_1}$ with length 1 on L towards Z_1. G walks along arc $\widehat{A_1B_1}$. Make a segment circle with radius $r_1 = \overline{OA_1}$, which intersects L with B_1. G walks along with arc $\widehat{A_1B_1}$. $j = j + 1, k = k + 1$.

Step 3:   While $k \geq 2$, do the followings:

(1) Make a segment $\overline{B_{k-1}A_k}$ with length $|r_{k-1}|$ on L towards Z_j. G walks along $\overline{B_{k-1}A_k}$. Make a semicircle with radius $r_k = \overline{OA_k}$, which intersects L with B_k. G walks along with arc $\widehat{A_kB_k}$.

(2) if (G reach the boundary of P) stop.

(3) $j = j + 1, k = k + 1$.

(4) if ($j > 2$) let $j = 1$.

We now pose several open problems for further research. Considering the spiral evacuation strategy, suppose that the evacuees basically move on a sequence of the semicircles, how could an efficient strategy be designed to solve the evacuation problem discussed in this paper? Also, it may be very interesting to study the evacuation problem with the spiral evacuation strategy, when $n = 2$. Moreover, we suggest that the spiral evacuation strategy is useful for solving the evacuation problems even when the affected area is not convex region.

Acknowledgements The authors would like to thank the anonymous referees for their valuable comments and suggestions for improvements. The work by Jiang was partially supported by National Natural Science Foundation of China under grant 61173034 and the General Project of Liaoning Province Science and Research (No. L2015105).

References

1. X. Chen, F.B. Zhan, Agent-based simulation of evacuation strategies under different road network structures. J. Oper. Res. Soc. **59**, 25–33 (2008)
2. Q.S. Lu, B. George, S. Shekhar, Capacity constrained routing algorithms for evacuation planning: a summary of results, in *Advances in Spatial and Temporal Databases*. Lecture Notes in Computer Science, vol. 3633 (Springer, Berlin, 2005), pp. 291–307

3. S. Shekhar, K. Yang, V.M.V. Gunturi, L. Manikonda, et al., Experiences with evacuation route planning algorithms. Int. J. Geogr. Inf. Sci. **26**(12), 2253–2265 (2012)
4. P. Berman, On-line searching and navigation, in *Competitive Analysis of Algorithms* (Springer, Berlin, 1998)
5. W. Burgard, M. Moors, D. Fox, R. Simmons, S. Thrum, Collaborative multi robot exploration, in *Proceedings 2000 ICRA. Millennium Conference*. IEEE International Conference on Robotics and Automation. Symposia Proceedings, vol. 1, pp. 476–481 (2000)
6. Y. Xu, L. Qin, Strategies of groups evacuation from a convex region in the plane, in *Frontiers in Algorithmics and Algorithmic Aspects in Information and Management*. Lecture Notes in Computer Science. vol. 7924 (Springer, Berlin, 2013), pp. 250–260
7. Q. Wei, L. Wang, B. Jiang, Tactics for evacuating from an affected area. ICCCI, IACSIT **3**(5), 435–439 (2013)
8. Q. Wei, X. Tan, B. Jiang, et al., On-line strategies for evacuating from a convex region in the plane, in *The Proceedings of Combinatorial Optimization and Applications*. Lecture Notes in Computer Science, vol. 74–85 (2014)
9. R.A. Baeza-Yates, J.C. Culberson, G.J.E. Rawlins, Searching in the plane. Inf. Sci. **106**(2), 234–252 (1993)

Chapter 12
A New Computational Model of Bigdata

Binhai Zhu

Abstract BigData has been a buzzword since a few years ago. However, what is exactly the corresponding (theoretical) computer model? What can be done and cannot be done with such a model? These are questions demanding answers. Recently, a model was proposed to address this issue by simulating a restricted version of the PRAM model. In this paper, we propose a theoretical model called Master/Slave Multiprocessor (MSM for short) which is very similar to a practical system using MapReduce but with additional constraints relevant to BigData processing. This model captures some of the most important properties of the practical coarse-grained multiprocessors (CGM) model (instead of PRAM). The most important ideas under such a master/slave model are that (1) even the master can only access a small fraction of all the data in its slave processors at any given time, (2) a large amount of data transmission between the slave nodes is considered impossible or too costly, and (3) an extra slave processor, together with the data it carries, can be easily integrated into the system to support scalability. Under such a model capturing the most important characteristics of a practical MapReduce system, some standard problems, like sorting, become hard to solve. We then propose an adaptive MSM model where the master node still has limited working memory but a large secondary storage. We demonstrate how these two MSM models can be used to solve some basic problems when the data volume is huge.

12.1 Introduction

BigData has been a buzzword since a few years ago. In almost every industry we have heard of people talking about BigData, one prominent example is the huge data volume of all Walmart transactions, or the search queries Google has to process, or a video surveillance system monitoring all the road intersections in New York City, all just in a single day. Funding agencies in different countries are setting up research

B. Zhu (✉)
Department of Computer Science, Montana State University, Bozeman, MT 59717, USA
e-mail: bhz@cs.montana.edu

© Springer International Publishing Switzerland 2015   201
L.M. Chen et al., *Mathematical Problems in Data Science*,
DOI 10.1007/978-3-319-25127-1_12

programs aiming at handling BigData in different areas. In short, BigData are there and will change our lives. We need to process them properly, if not, they will become our burden.

Nevertheless, there does not seem to exist a commonly accepted model for BigData computations. As some level of parallelism must be used to handle BigData, let us first briefly review the known parallel computation models (see also [2]).

The most popular parallel model is PRAM (Parallel Random Access Machine) which was proposed in 1970s, but unfortunately a generic practical PRAM has never been built, though there has been some recent effort to do that [17]. On the other hand, there has been a lot of algorithmic research based on PRAM, see [11, 13]. But the gap between theoretical algorithms and practically working systems on PRAM remains huge. Seeing this gap, in early 1990s, the LogP [3] and BSP (Bulk Synchronous Parallel) models [16] were proposed, in essence, to have a distributed-memory, overall asynchronous parallel machine. In the early 1990s, the coarse-grained multiprocessors (CGM), which can be thought of as the restricted version of LogP and BSP, are proposed. In short, a CGM is a set of processors, each with enough local memory and sufficient computation power, and the synchronization is through point-to-point communication. In fact, two practical systems, PVM—parallel virtual machine, and MPI—message passing interface, are implemented and used widely by professionals in parallel computation. In fact, two special issues on CGM algorithms have been published in scientific journals [5, 6]. However, CGM require high communication bandwidth and system connectivity for real-time applications, hence are either too expensive or are not really practically applicable for many commercial applications, especially when the data volume is huge and when a fully connected network is not always guaranteed.

Surprisingly, a master/slave model, called MapReduce, was adopted in late 1990s by Google and becomes almost prevalent in large scale web search [4]. The idea of MapReduce is as follows. (1) A master node maps and divides a task into several parts and assign them to its slave nodes (the Map process), and (2) the slave nodes finish the sub-tasks and return the results for the master node to combine for further computation (the Reduce process). Note that this process can be done in multiple rounds and can be done recursively, i.e., a slave node can divide the task further to a set of sub-slave nodes. Under this model, the communication is only between a master node and its slaves, the slaves cannot communicate directly among themselves. In Fig. 12.1, P_1 is the master node for slave nodes P_2, P_3, and P_4; and recursively P_4 is the master node for P_5 and P_6.

One of the most significant properties of the MapReduce system, from the application perspective, is the scalability. Namely, when one has an extra slave node available (together with some associated data), it is easy to add it under an existing master without changing the overall topology of the system. In Fig. 12.1, when node P_7 is added as a slave node under P_4, all other nodes, except P_4, are not affected. For CGM, one would have to add a communication channel between the new node to all the existing nodes, which is much more costly to implement.

It remains to tell what can and cannot be done efficiently using a MapReduce system. The practical success on MapReduce is mainly on the searching and

Fig. 12.1 Illustration of a
practical MapReduce system

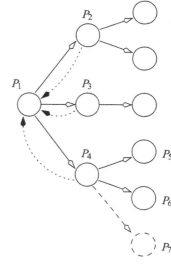

Fig. 12.2 The master/slave
multiprocessor model

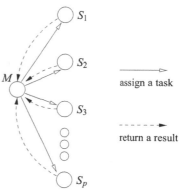

querying side (see [10] for a most recent survey). With this motivation, in 2010, Karloff, Suri, and Vassilvitskii proposed a MapReduce model, which is based on simulating a restricted version of PRAM [12]. Under this model, many graph problems can be solved efficiently [14]. However, as this model inherits some of the drawbacks of PRAM, we believe that it is probably too powerful (or practically too costly) compared with any existing practical MapReduce system.

Intending to obtain a system based on LogP and BSP, instead of PRAM, in this paper, we propose a Master/Slave Multiprocessor model (MSM) based on MapReduce to handle large data sets (or BigData). Abstracted from some existing practical system using the MapReduce model, we assume that this MSM has a master node M and p slave nodes $S_1, S_2, \ldots, S_p$; moreover, the internal memory of the master node M is in the same order as those slaves nodes. (In practice, that means it is easy to replace the master node when it becomes faulty.) To be able to analyze the efficiency of algorithms, we assume that each S_i and M has a memory of size $O(\frac{n}{p})$, where n is the input size. Note that here p should not be taken as a small constant. See Fig. 12.2 for an example.

This paper is organized as follows. In Sect. 12.2 we introduce the model. In Sect. 12.3 we show how to solve some fundamental problems using this model and also leave some open problems. Moreover, we show an adaptive version of this model. In Sect. 12.4, we conclude the paper.

12.2   MSM: The Master/Slave Multiprocessor Model

In this section, we discuss some technical aspects of the MSM model, especially from the processing of BigData point of view.

The measurement of algorithm complexity under the MSM model includes

- #C: The number of communication rounds, where each round includes a ⟨master, slave⟩ and ⟨slave, master⟩ communication, possibly including sending $O(\frac{n}{p})$ amount of data, where n is the input size and p is the number of slave nodes.
- T: The local computation cost within one round of communication.
- D: The maximum amount of information exchanged between the master and slave nodes within one round of communication.

The computation complexity (cost) of the algorithm is $O(\#C \cdot T)$. The total communication cost of the algorithm is then $O(\#C \cdot p \cdot D)$.

Different from some practical MapReduce system, like Hadoop, we assume that the master node M and all the slave nodes have a local internal memory of $O(\frac{n}{p})$. This assumption is valid when the data sets are large, for instance, all the data in Google cannot be stored in any single computer. We assume throughout this paper that $n > p^2$. Also, in practice communication cost is high; hence, it is likely that a practical algorithm only allows a small (preferably constant) number of communication rounds, and within each round a limited amount of information is exchanged. The last constraint implies that it is hard to swap a large amount of data between a master node and a slave node. In fact, without this constraint, one can always simulate a CGM algorithm with an MSM, as the communication between two (slave) nodes can be done through the master node.

Observation 1. *If there is no constraint on communication, then a CGM algorithm can be simulated with an MSM system where a point-to-point communication can be simulated with at most a pair of ⟨slave,master⟩ and ⟨master,slave⟩ communications, and a broadcast communication can be simulated with at most one ⟨slave,master⟩ communication and one round of ⟨master,slave⟩ communications.*

From now on, we forbid the data transmissions between slave nodes. Under this assumption (also the limited working memory of the master node), we notice that many standard problems become hard to solve, for example, sorting n elements stored in p processors. The reason is that the Reduce step needs to send the p partially sorted data items back to the master node, which is impossible unless we can store all of the n data items in the master node or we allow a large amount of information exchange. That gives us the following observation.

Observation 2. *The MSM system can efficiently solve a problem of size n where there are a bounded number of communication rounds and in each round the problem can be partitioned into p independent parts and each results in a sub-solution of size $O(\frac{n}{p^2})$.*

The above observation basically shows that a naive simulation of many algorithms on an MSM is not going to work. For instance, one can solve the diameter problem in any fixed dimension in $O(n^2)$ time with a brute-force method. There are n rounds of computations/communications: each round the master asks the slave nodes to compute the maximum distance to a fixed point and always keeps the current maximum. Of course, the number of rounds, n, would be too large.

However, some problems can be solved efficiently using this MSM model. For example, a search on Google can be best simulated as a $\leq k$-nearest neighbor search problem on an MSM (where k is typically a small constant), with only one round of communication: The master node assigns the search to p slave nodes, each returns the $\leq k$-nearest neighbors at each slave node. These neighbors are then returned to the master node so that it can combine these results to compute the $\leq k$-nearest neighbors in the whole system, according to the distance between the query object and the searched objects. Note that at this stage some results could be truncated if the returned p sets of partial results are too large.

In the next section, we try to solve several basic (geometric) problems using this MSM model.

12.3   Processing Large Data sets with an MSM

In this section, we illustrate how to use the MSM model to solve a few basic problems when the data volume (i.e., n) is huge. (We try to keep the paper self-contained, but readers can refer to [15] for anything related to computational geometry.) We assume that the data are stored roughly in equal size in each slave node; however, we do not assume any other property of the stored data, e.g., locally distributed or organized into some data structure—the latter can be rectified in practice, as building some complex data structure locally is not much different from regular computations.

This assumption is critical for the remaining parts of this paper, and we believe that it is practical as in practice the data sets might be generated, maintained, and updated in different ways and stored locally. In fact, if we allow the shuffling of the input data sets in an arbitrary fashion, many problems can be solved easily by tuning the existing CGM algorithms. For instance, if we assume that the master can generate a random sample and use the sample to partition and map the remaining data to the corresponding slave nodes, we can then solve many problems more efficiently with random sampling by tuning the existing CGM algorithms, for instance, 3D convex hull (2D Voronoi diagram) [7], or 2D Voronoi diagram for line segments [9]. In fact, even some CGM algorithms on graph problems [8] could be simulated on such an MSM model with large communication overhead.

However, we think that this does not fit most applications in processing BigData, where in many situations data are generated and stored first, but we do not know what we want to compute from the data until later on. For practical consideration, we only allow a small number of communication rounds and try to bound the communication costs as much as we could. Due to this reason, for most problems under consideration, we can only design approximation algorithms.

12.3.1 Selection

It is easy to see that with one round of communication the Min, Max of n 1-D values X stored in an MSM can be computed in $O(\frac{n}{p})$ time. Each slave node just needs to perform the selection of the extreme (Min or Max) values and the $\frac{n}{p}$ returned values can be combined at the master node to obtain the overall extreme (Min or Max) value for the set X.

We next show that the selection of the k-th smallest value of X can be solved efficiently on an MSM. Here k is not necessarily a small constant—otherwise the problem will be easy to solve.

Theorem 12.1. *Given a set X of n 1-D values, with $\min\{\log R, N\}$ rounds of communication and a total of $O(\min\{\log R, N\} \cdot p)$ communication costs, the k-th smallest element of X can be computed in $O(\frac{n}{p} \cdot \min\{\log R, N\})$ time on an MSM with p slave processors, where R is the distance between $\min_1$ and $\max_1$, $N \leq n(1 - 1/p)$ is the number of elements between $\min_1$ and $\max_1$, $\min_1$ is the minimum element in X and $\max_1$ is the minimum k-smallest elements in the p slave nodes, respectively.*

Proof. The local computation cost at step 2 is obviously $O(\frac{n}{p})$ [1]. The number of communication rounds is first bounded by $O(\log R)$ and in the worst case could be

---

Algorithm 12.1

---

1. M computes the smallest value *min* in set X.
2. M assigns the task of finding the k-th smallest value to all the slave nodes $S_1, \ldots, S_p$.
3. Each S_i returns the k-th smallest value $x_i^k, i \in [1, p]$, to M.
4. M selects the minimum value among the p returned values from the slave nodes. Let the value be max.
5. Then, M computes $L = (\max + \min)/2$ and assigns the task to the slave nodes whether there are k elements in X which are $\leq L$. Each slave node performs this task by counting locally the number of values $\leq L$. If there are exactly k elements in X which are $\leq L$, then with one extra round of communication, M can find and return the largest value in X which is $\leq L$. If there are $< k$ elements in X which are $\leq L$, then compute the largest element in X which is $\leq L$, still call it min, update $L \leftarrow (\max + \min)/2$. If there are $> k$ elements in X which are $\leq L$, then compute the smallest element in X which is $> L$, still call it max, update $L \leftarrow (\max + \min)/2$. Repeat until the solution is returned.

$O(N)$. In fact, we can construct a set of points between $\min_1$ and $\max_1$ such that the interdistances grow exponentially, and hence the algorithm will hit every point for an extra round of communication.

For practical data, say the data are uniformly distributed or uniformly distributed in a constant number of clusters, it is safe to claim that $O(\log R) = O(\log n)$.

Open Problem 1. For the selection problem, is it possible to design an algorithm under the MSM with a local computation cost of $O(\frac{n}{p})$ and with $O(\log n)$ rounds of communications?

12.3.2 Diameter in any Fixed Dimension

Given a set P of n points in any fixed dimension, the diameter is the maximum distance between two points in P. As discussed before, if we allow n rounds of communications, then the problem can be easily solved optimally. We discuss how to design algorithms with few rounds of communications.

Theorem 12.2. *Given a set P of n points in any fixed dimension, with one round of communication and total of $O(p)$ communication cost, a 2-approximation of the diameter of P can be computed in $O(\frac{n}{p})$ time on an MSM with p slave processors.*

Proof. Let $d(u, v)$ be the diameter. By triangle inequality, $d(u, v) \leq d(u, q) + d(q, v)$. Then, if the algorithm selects the farthest point z from q, we have $d(z, q) \geq \max\{d(u, q), d(q, v)\} \geq \frac{1}{2}d(u, v)$.

We comment that this algorithm is from a folklore diameter algorithm and could certainly be improved. For instance, let z be the point returned by the algorithm, we can run the algorithm one more time to find the farther point y from z, then $d(z, y)$ would be a factor-$\sqrt{3}$ approximation for the diameter. Here, the purpose is just to show the application of the MSM model.

12.3.3 Sorting and Element Uniqueness

Under the constraint that even the master node can only store $O(\frac{n}{p})$ amount of data. Sorting becomes a hard problem on an MSM. In fact, even a related Element

---

Algorithm 12.2

---
1. M picks any point $q = p_{s,t}$ and assign the task of finding the maximum distance from q to all the slave nodes $S_1, \ldots, S_p$.
2. Each S_i returns the farthest point $p_{i,i'}$, $i \in [1, p]$, $i' \in [1, \frac{n}{p}]$, to M.
3. M selects the farthest point from q among the p returned points from the slave nodes.

Uniqueness problem, namely, determining whether there are two equal elements (possibly stored in different slave nodes), becomes hard. We leave this as an open problem.

Open Problem 2. For the element uniqueness problem, is there an efficient solution under the MSM?

12.3.4 Relaxing the Master Memory Constraint in an MSM

As we discussed before, if the master node in an MSM has only a size of $O(\frac{n}{p})$ then even the sorting problem becomes difficult. On the other hand, if the size of the master node has a memory of size $O(n)$ (which is roughly the sizes of all its slave nodes), then the model is not realistic (though many problems can be efficiently solved). Here, we try to make the model slightly adaptive. We assume that the master node has a (working) memory of $O(\frac{n}{p})$, but with some extra cost a secondary storage can be expanded to the master node—to store the eventual computational results. We next illustrate a simple example of solving the sorting problem under this adaptive model. We comment that this adaptive model is still different from a practical MapReduce model, like Hadoop, where the (working) memory of the master node could be much larger than $O(\frac{n}{p})$.

If the master node has a secondary storage which is of size $O(n)$, then sorting becomes solvable—even if the working memory at the master level is still $O(\frac{n}{p})$. We can sort all elements stored in each slave node and then each node sends the first $\frac{n}{p^2}$ smallest elements to the master node. Among the p lists of $\frac{n}{p^2}$ elements (for a total of $\frac{n}{p}$) received, the master node can extract elements from the p lists, compute and dump the smallest elements in a buffer. If one of the lists, L_i from S_i, is empty, then S_i sends its current $\frac{n}{p^2}$ smallest elements to L_i at M, and the buffer contents are written to the secondary storage and then emptied. Then the process can be repeated. Note that the buffer is of size $\Omega(\frac{n}{p^2})$.

Theorem 12.3. *Given a set X of n 1-D values, with $O(p^2)$ rounds of communication and a total of $O(n)$ communication costs, the elements of X can be sorted in $O(\frac{n \log n}{p})$ time at each of the p slave processors of an MSM, and it takes the master node $O(n)$ time to compute and store the sorted list of X in a secondary storage.*

Proof. It is easy to see that there are at most $p^2 - p = O(p^2)$ rounds of communication. The computation time of the master node is $O(n - \frac{n}{p}) = O(n)$. $\blacksquare$

We comment that the above algorithm is a naive implementation of Merge–Sort and it would be interesting to see an improved algorithm.

12.4 Concluding Remarks

In this paper, we try to propose a theoretical model capturing the most important properties of BigData computations. The model we propose is a master/slave multiprocessor incorporating all important properties of a MapReduce system. One important constraint we put on this MSM model is that the master node cannot access all the input data at the same time and large amount of data transmission between nodes is forbidden. This might be the reason that some fundamental problems, like sorting, becomes hard to solve. We also propose an adaptive model where the master node has limited working memory but a much larger secondary storage. We show some simple applications of both of the models.

Acknowledgements This research is partially supported by NSF of China under grant 60928006. Thanks to Dr. Li Chen for organizing the ICM'2014 Workshop on Mathematical Foundation of Modern Data Sciences, where this work was initiated.

References

1. M. Blum, R. Floyd, V. Pratt, R. Rivest, R. Tarjan, Time bounds for selection. J. Comput. Syst. Sci. **7**(4), 448–461 (1973)
2. D. Campbell, A survey of models of parallel computation. Technical Report, University of York (1997)
3. D. Culler, R. Karp, D. Patterson, A. Sahay, K. Schauser, E. Santos, R. Subramonian, T. von Eicken, LogP: towards a realistic model of parallel computation, in *ACM SIGPLAN Symposium on Principles and Practice of Parallel Programming*, pp. 1–12 (1993)
4. J. Dean, S. Ghemawat, MapReduce: simplified data processing on large clusters, in *Proceedings of the OSDI*, pp. 137–150 (2004)
5. F. Dehne, Guest Editor's Introduction, special issue on coarse grained parallel algorithms. Algorithmica **24**(3–4), 173–176 (1999)
6. F. Dehne, Guest Editor's Introduction, special issue on coarse grained parallel algorithms for scientific applications. Algorithmica **45**(3), 263–267 (2006)
7. F. Dehne, X. Deng, P. Dymond, A. Fabri, A. Khokhar, A randomized parallel three-dimensional convex hull algorithm for coarse-grained multicomputers. Theory Comput. Syst. **30**(6), 547–558 (1997)
8. F. Dehne, A. Ferreira, E. Caceres, S.W. Song, A. Roncato, Efficient parallel graph algorithms for coarse-grained multicomputers and BSP. Algorithmica **33**(2), 183–200 (2002)
9. X. Deng, B. Zhu, A randomized algorithm for the Voronoi diagram of line segments on coarse-grained multiprocessors. Algorithmica **24**(3–4), 270–286 (1999)
10. W. Fan, J. Huai, Querying big data: bridging theory and practice. J. Comput. Sci. Technol. **29**(5), 849–869 (2014)
11. J. JaJa, *An Introduction to Parallel Algorithms* (Addison-Wesley, Reading, 1992)
12. H. Karloff, S. Suri, S. Vassilvitskii, A model of computation for MapReduce, in *Proceedings of the 21st Annual ACM-SIAM Symposium Discrete Algorithms*, pp. 938–948 (2010)
13. R. Karp, V. Ramachandran, A survey of parallel algorithms for shared-memory machines. Technical Report UCB/CSD-88-408, University of California-Berkeley (1988)
14. S. Lattanzi, B. Mosley, S. Suri, S. Vassilvitskii, Filtering: a method for solving graph problems in MapReduce, in *Proceedings of the 22nd Annual ACM Symposium on Parallelism in Algorithms and Architectures*, pp. 85–94 (2011)

15. F. Preparata, M. Shamos, *Computational Geometry* (Springer, New York, 1985)
16. L. Valiant, A bridging model for parallel computation. Commun. ACM **33**(8), 103–111 (1990)
17. U. Vishkin, Using simple abstraction to reinvent computing for parallelism. Commun. ACM **54**(1), 75–85 (2011)

Index

A

Algorithmic geometry, 103–108
Amazon AWS, 12
Approximation, 23, 26, 31, 33, 40, 49–52,
 69–71, 76, 88, 120, 126, 131, 134–138,
 144–147, 172, 177–179, 181, 182, 186,
 206, 207
Artificial intelligence (AI), 3, 6, 7, 17, 18, 23,
 35, 40, 47, 72

B

Bayesian network, 72, 73
Betti number, 110, 111, 113–116, 118
Bhattacharyya coefficient, 97
BigData, vii, viii, 3–14, 24, 31, 34, 39, 56,
 57, 71–73, 75–99, 108, 117, 121, 122,
 201–209
Big pixel, 43, 81, 88–90, 121
Boosting method, 64
Bounding box, 95, 96, 98
Breadth first search, 18, 20, 21, 47, 56, 121

C

Classification, vii, 3, 4, 9, 18–19, 25–26, 41,
 42, 48, 53, 56, 64–66, 69–71, 76, 77,
 79, 80, 89, 101, 115, 120–122, 148,
 150–152, 154
Cloud computing, vii, 3–5, 7, 8, 10, 12, 14,
 31, 34, 35, 40, 76, 77, 88, 89, 102, 108,
 120–123
Cloud data, 8–10, 55, 102–103, 105–109, 112,
 118
Clustering, 3–5, 11, 25–26, 46, 73, 79, 80,
 89–93, 120, 176, 207

Computational geometry, 103, 205
Computational learning, 70–71
Computational topology, 104
Connected graph, 5
Connectedness, vii, 39–46, 48–57, 72, 81–87,
 90, 119–122
Connectivity, vii, viii, 4, 5, 8, 13, 14, 20–22,
 34, 39–57, 64, 72, 73, 80–87, 89–92,
 99, 109–111, 113, 115, 116, 119–122,
 197, 202
Convex polygon, 105, 190, 193, 194
Correlation, 8, 24, 96, 129, 130
Covariance, 23
Covariance matrix, 24

D

Data fitting, 26, 27, 49–52
Data mining, 3–4, 6, 7, 12, 73, 90, 102
Data reconstruction, 9, 26–27, 49–52, 55, 56,
 119, 121
Data recovery, vii, 8, 10, 49
Data science, vii, viii, 3–14, 17–35, 39–42,
 56, 57, 63–73, 76, 108, 117, 121, 122,
 131–138
Data structure, vii, 21, 28–29, 34, 48, 108, 205
Decision tree, 9, 18, 63–64, 66, 71, 79, 97
Decomposition, 11, 29, 43, 46–49, 56, 93,
 102–104, 122, 137
Delaunay triangulation, 103–105
Digital geometry, 113
Digital topology, 113–116
Dijkstra's algorithm, 22, 107, 108
Dimension reduction, vii, 10, 35, 92, 106
Directed graph, 32, 42–45, 50, 72

Printed in the United States
By Bookmasters